ELECTRONICS ENGINEERING FOR PROFESSIONAL ENGINEERS' EXAMINATIONS

Other McGraw-Hill Reference Books of Interest

Professional Engineering Books

Constance • ELECTRICAL ENGINEERING FOR PROFESSIONAL ENGINEERS' EXAMINATIONS

Constance • HOW TO BECOME A PROFESSIONAL ENGINEER

Constance • MECHANICAL ENGINEERING FOR PROFESSIONAL ENGINEERS' EXAMINATIONS

Kurtz • ENGINEERING ECONOMICS FOR PROFESSIONAL ENGINEERS' EXAMINATIONS

Kurtz • STRUCTURAL ENGINEERING FOR PROFESSIONAL ENGINEERS' EXAMINATIONS: Including Statistics, Mechanics of Materials, and Civil Engineering

LaLonde, Jr. and Stack-Staikidis • PROFESSIONAL ENGINEER'S EXAMINATION QUESTIONS AND ANSWERS

Polenz • ENGINEERING FUNDAMENTALS FOR PROFESSIONAL ENGINEERS' EXAMINATIONS

Prabhudesai and Das • CHEMICAL ENGINEERING FOR PROFESSIONAL ENGINEERS' EXAMINATIONS

For more information about other McGraw-Hill materials, call 1-800-2-MCGRAW in the United States. In other countries, call your nearest McGraw-Hill office.

ELECTRONICS ENGINEERING FOR PROFESSIONAL ENGINEERS' EXAMINATIONS

Charles R. Hafer, P.E.

Second Edition

McGraw-Hill Publishing Company

New York   St. Louis   San Francisco   Auckland   Bogotá
Caracas   Hamburg   Lisbon   London   Madrid   Mexico
Milan   Montreal   New Delhi   Oklahoma City
Paris   San Juan   São Paulo   Singapore
Sydney   Tokyo   Toronto

To:
My mother, Evolena; my father, Charles W.;
my wife, Medda; and my son, Mark.

Library of Congress Cataloging-in-Publication Data

Hafer, Charles R., date.
  Electronics engineering for professional engineers'
examinations/
Charles R. Hafer.—2nd ed.
    p. cm.
  Bibliography: p.
  Includes index.
  ISBN 0-07-025433-8
  1. Electronics—Examinations, questions, etc.
2. Electronics—Problems, exercises, etc.   I. Title.
TK7863.H33   1989
621.381'076—dc19   88-35689
                                                         CIP

1234567890   DOC/DOC   895432109

ISBN 0-07-025433-8

The editors for this book were Daniel A. Gonneau and Nancy Young,
and the production supervisor was Richard A. Ausburn.

For more information about other McGraw-Hill materials,
call 1-800-2-MCGRAW in the United States. In other
countries, call your nearest McGraw-Hill office.

CONTENTS

Preface   vi

Acknowledgments   ix

Symbols and Abbreviations   x

Chapter 1   Discrete Semiconductor and Integrated
Circuit Problems   1

Chapter 2   Circuit Analysis Problems   77

Chapter 3   Servo Control and Feedback Problems   125

Chapter 4   Instrumentation, Computer, and System
Problems   177

Appendix A.   Bipolar Transistor Gain and Impedance
Equations   241

Appendix B.   JFET Small-Signal Equations   256

Appendix C.   Operational Amplifier Equations   262

Appendix D.   Miscellaneous Circuit Equations   269

Appendix E.   Laplace Transform Table of Forcing
Functions   275

Appendix F.   JFET DC and AC Parameter Equations   277

Appendix G.   Transfer Function Plots for Typical
Transfer Functions   289

Appendix H.   Standard Component Values   298

Appendix I.   Fourier Series Waveform Equations   300

Appendix J.   RC Network Synthesis for Feedback
Amplifiers   302

Appendix K.   RC Transfer Function Equations   319

Bibliography,   **330**

Index,   **332**

PREFACE

At the time the first edition of this book was being written, much of the material available for preparing for the Principles and Practices portion of the National Council of Engineering Examiners (NCEE) professional engineering licensing examination was aimed at helping the electrical engineer who practiced commercial power engineering or industrial electrical engineering. Since the material available to aid the practicing electronics engineer was limited, this book was written to fill that void. This edition was created to include some of the more recent types of problems and to eliminate those that were considered obsolete.

The book includes problems and solutions similar to those on the NCEE examination and provides reference material, in the form of appendices, which can be used in support of these solutions. All states have adopted the NCEE examination for their electrical engineering licensing examinations, and the material presented in this book is oriented toward that examination.

The NCEE test is intended to be a comprehensive and fair examination designed to test the ability of the applicant to practice as a professional electrical engineer. There is never any guaranteed format for the selection of the problems, but information provided by the NCEE indicates that approximately 65 percent of the problems are on electronics and 35 percent on power (including industrial and commercial power). Usually there are 24 problems, of which the candidate must work 8. A passing score is normally set at 70 percent. The following is a typical breakdown of the problems given:

	Morning	Afternoon
Power and machinery	4	4
Electronics	3	3
Computers	2	2
Controls	3	2
Economics		1
Total	12	12

The intent of this book is to cover only the electronics subject areas except economics. Economics is important in all engineering disciplines, but has been thoroughly treated in many other publications.

The subjects comprising the power portion are power and systems, machines, circuits, economics, and illumination. The subjects considered as electronic are electronics, computers, and controls.

Although the NCEE examination has been adopted by all states, the method of administering the examination and additional requirements is at the discretion of the particular state board. For example, the test may be closed or open book; a personal interview may be required; the usable reference material may be limited; and qualifications may vary. The requirements for the P.E. licensing examination usually include 4 years of applicable experience from the date of graduation for a degreed engineer from an accredited institution, and 10 years for the non-degreed. For persons with related 4-year degrees, such as in math or physics, 6 years of experience are normally required for qualification. The Engineering-In-Training (EIT) examination is a prerequisite and may be taken at the same examination time frame as the Principles and Practice portion or any time earlier at the discretion of the state board. However, a separation of the examinations eases the quantity of material to be studied at one time.

Several general guidelines should be observed when preparing for the examination. In particular, the following is suggested:

1. File for your license early so that your application can be processed in a timely fashion. A minimum of 3 months before the examination is suggested, with 6 months providing a more comfortable margin. (Tests are given in April and October as of 1989.)

2. Start studying at least three months before examination time. Working problems provides an objective approach that is a proven technique with most technical subjects. Budget a minimum of 250 hours for study. If nothing else, this study will help formulate your thoughts and approaches to various problems.

3. During your study period, create a "things to take/do" list. One such list might include:

Take tables, ruler, compass, erasers, graph and plain paper, two boxes of sharpened pencils, two fully charged identical calculators (preprogrammed calculators are usually not allowed), multi-outlet extension cord, books (list), references (list), handcart (to carry everything), aspirin. Index your notes and books.

4. Be sure to have the exact examination location and if possible make a visit to the site before the final day. Investigate the eating facilities so you know whether or not to take a lunch.

5. Leave in plenty of time for arrival the day of the examination.

6. If the test is out of town, arrive the day before, sleep there overnight, and get plenty of rest.

7. Don't study the last night; if you need to, you're not ready.

Eight hours is normally given for the test and it is divided into two sessions: morning and afternoon. In each session, usually 12 problems are given, from which you must work four. In some states, you may receive all of all the problems in the morning and may finish your eight problems in any order, as long as you pick four from each session. Follow instructions and turn in your working paper. Work the problems as neatly and quickly as possible and show all of the work. After you have completed your problems, check them if you have the time. Partial credit is given. Above all, pay attention to the instructions. As an example, the instructions might require an economics problem to be worked. If you use a reference, list it in your solution area and include the title, author, page, and edition. This is important!

The material presented in this book assumes that the user has a minimum of a bachelor's degree, or equivalent experience, because of the mathematics and electronic analysis backgrounds required. The material presented may be used as instructional material, but is primarily intended for review.

This book is separated into four chapters. Chapter 1 contains problems dealing with semiconductors, such as bipolar transistors, diodes, JFETs, and integrated circuits. Chapter 2 contains problems dealing with ac, dc, and transient circuits; and operational amplifiers proving circuit concepts. Chapter 3 contains problems dealing with control systems, while Chapter 4 deals with problems involving instrumentation, computers, and systems.

Do not be alarmed at the solution length of some of the problems given in the text, because those given on the examination can usually be done on two or fewer handwritten pages. Many problems in the text contain the equivalent of more than two pages of handwritten solutions, but this is the result of the tutorial-type presentation of the material or the increased material presented to give the candidate more problem-solution practice.

It is recommended that the candidate become familiar with the material presented in the Appendices as it can be extremely helpful in the problem solutions and can be used as a reference on the examination. If the problems are found to be too difficult, it may be necessary to review some texts that are basic in the covered areas. Some texts that may prove useful for review and as references are given in the Bibliography.

Charles R. Hafer

ACKNOWLEDGMENTS

I consider it a privilege to acknowledge those people who assisted me in the preparation of this text. I sincerely wish to thank Madjid Belkerdid, Basile Dimitriatis, Bill Herzog, and Dan Williams for their inputs and solutions checks. Special thanks go to two of my colleagues, Jack Burton and Jim Logie. Jack worked with me for more than a year on the original manuscript and checked every problem as well as the Appendices. Jim reworked every problem at the galley proof stage.

Most importantly, my sincere appreciation and thanks go to my dear wife Medda, whose encouragement, patience, and understanding were instrumental in the successful completion of this book. She typed the original manuscript over a 9-month period and spent several months of manuscript checking at various stages of the text's completion.

During the first revision of this text, I'd like to thank Charles Maynard for his timely solution checks to the new problems.

SYMBOLS AND ABBREVIATIONS

$\downarrow$	Symbol for ground
$\parallel$	In parallel with
α	DC current gain, I_C/I_E
β	h_{FE}; dc current gain, I_C/I_B
Δ	Incremental change
Δh	$h_i h_o - h_r h_f$
η	Efficiency
ϕ_m	Phase margin
ω	Angular frequency, $2\pi f$
Ω	Ohms
A_V	Voltage gain
A_I	Current gain
CD, CG, CS	JFET amplifier configurations: common drain, common gate, and common source
CR	Diode designator
$C\pi, C\mu$	Capacitor designations used in hybrid-π model
CB, CC, CE	Transistor configurations: common base, common collector, common emitter
f_H	Upper 3 dB frequency
f_L	Lower 3 dB frequency
f_M	Midband frequency
$G(s)$	Gain expressed in Laplace notation
$G(j\omega)$	Gain expressed as a function of ω
$G(s)H(s)$	Loop gain expressed in Laplace notation
$GH(j\omega)$	Loop gain expressed as a function of ω
g_{fs}, g_{fso}, g_o	Small-signal JFET parameters
g_m	Hybrid-π small-signal bipolar transistor parameter $(I_E/0.026)$
$h_{ib}, h_{ob}, h_{fb}, h_{rb}$	Small-signal common-base transistor parameters
$h_{ic}, h_{oc}, h_{fc}, h_{rc}$	Small-signal common-collector transistor parameters
$h_{ie}, h_{oe}, h_{fe}, h_{re}$	Small-signal common-emitter transistor parameters
h_{fe}	DC transistor current gain; h_{FE}; I_C/I_B

I_B	Bipolar dc base current
I_C	Bipolar dc collector current
I_D	JFET dc drain current
I_E	Bipolar dc emitter current
I_{DSS}	JFET dc saturation drain current
I_{GSS}	JFET gate leakage current
I_E	Bipolar dc emitter current
I_G	JFET dc gate current
I_S	JFET dc source current
JFET	Junction field-effect transistor
K	Boltzman's constant $(1.3 \times 10^{-23}$ J/K)
KT/q	Intrinsic transistor parameter; equals 0.026 V^1 at 25°C
r_d	Small-signal diode ac resistance
r_{DS}	JFET on resistance in triode region
r_{DSO}	Minimum value of r_{DS} occurring at zero values for V_{GS} and V_{DS}
r_o, r_x, r_π	Bipolar hybrid-π representation for resistance parameters
S	Siemens (formerly mhos)
S_I	Current stability factor
S_V	Voltage stability factor
V_B	Bipolar dc base voltage
V_{BE}	Bipolar dc base-emitter voltage
V_C	Bipolar dc collector voltage
V_{CC}, V_{DD}, V_{EE}, . . .	DC power supply voltages
V_D	JFET dc drain voltage
V_{DS}	JFET dc drain-to-source voltage
V_E	Bipolar dc emitter voltage
V_G	JFET dc gate voltage
V_{GS}	JFET dc gate-to-source voltage
V_{GS} (off)	JFET gate-to-source voltage required to reduce I_D to zero; sometimes referred to as pinchoff voltage
VR	Reference diode designation
V_P	JFET pinchoff voltage
V_S	JFET dc source voltage
V_{TH}	Thevenin's equivalent voltage
X_C	Capacitance reactance, $1/(2\pi f C)$
X_L	Inductance reactance, $2\pi f L$
Z_{IN}	Input impedance
Z_o	Output impedance
Z_T	Transfer impedance, e_{in}/i_o
Z_{TH}	Thevenin's equivalent impedance

ABOUT THE AUTHOR

Charles R. Hafer is an Engineering Manager with Martin Marietta Aerospace Corporation and has successfully engineered projects involving computers, guidance and control, avionics, electro-optics, and automatic-test systems. He has a BEE from the University of Florida and an MSEE from the Florida Institute of Technology.

1

DISCRETE SEMICONDUCTOR AND INTEGRATED CIRCUIT PROBLEMS

The problems contained in Chap. 1 deal with discrete semi-conductors and integrated circuits. Problems involving diode analysis, dc and ac transistor analysis, and integrated circuit analysis are included for review. Appendices A to F, H, J, and K will prove particularly useful in solving the problems in this chapter.

Although most of the problems in this chapter can be worked from equations given in the appendices, a good transistor/diode analysis book, a good JFET analysis book, and an operational amplifier analysis book should be taken as references to the examination.

Equations are represented in the text as follows: (F1-4) indicates the equation used comes from Appendix F1, and is equation 4.

PROBLEM 1

GIVEN: The circuit shown in Fig. 1-1 with the following element values:

$$R_s = 1 \text{ k}\Omega \quad R_3 = 4.7 \text{ k}\Omega \quad R_6 = 10 \text{ k}\Omega$$

$$R_1 = 47 \text{ k}\Omega \quad R_4 = 1 \text{ k}\Omega \quad R_E = 1 \text{ k}\Omega$$

$$R_2 = 10 \text{ k}\Omega \quad R_5 = 10 \text{ k}\Omega \quad R_C = 4.7 \text{ k}\Omega$$

$$h_{fe_1} = h_{fe_2} = 50 \quad\quad\quad V_{CC} = +20 \text{ V dc}$$

1

FIND:

   a.   Identify the amplifier configuration.

   b. $e_o/e_s = A_v$.

   c. $i_2/i_1 = A_i$.

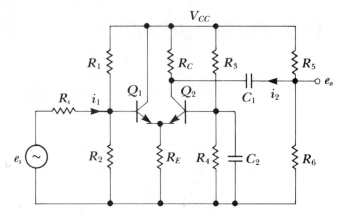

Fig. 1-1   Amplifier configuration for Prob. 1.

SOLUTION:

   a. ˙ Since this is all the information given, X_{C1} and X_{C2} will be assumed to be zero impedances at the frequency of interest. Also, since no parameter values were given, we will assume approximate analysis will be sufficient. We could estimate all parameters, but this would not be feasible because the solution would become quite cumbersome since the amplifier is multistage. Trace the signal path from input to output as shown in Fig. 1-2. This shows a common-collector (CC) amplifier driving a common-base (CB) amplifier. Let's draw the ac equivalent circuit taking all power supplies to ac ground and making all capacitors short circuits. The ac equivalent circuit is shown in Fig. 1-3. The linear ac equivalent circuit is shown in Fig. 1-4. The equations that describe the CC and CB amplifiers are given in Appendices A3 and A1, respectively.

   b.   Although we are able to use approximate gain equations, we need to estimate the emitter currents so we can find $h_{ib,1}$ and $h_{ib,2}$. We will assume Q_1 and Q_2 are matched, $I_{E1} = I_{E2}$, and e_s is ac-coupled.

      Solve for I_{E1} and I_{E2}:

$$V_{B2} \approx \frac{R_4 V_{CC}}{R_3 + R_4} = \frac{1000(20)}{4700 + 1000} = 3.51 \text{ V dc}$$

$$V_{E2} \approx V_{B2} - V_{BE2} \qquad \text{(assume } V_{BE2} = 0.7 \text{ V dc)}$$

$$V_{E2} = 3.51 - 0.7 = 2.81 \text{ V dc}$$

$$I_{E1} = I_{E2} = \left(\frac{2.81}{1000}\right)\frac{1}{2} = 1.405 \text{ mA}$$

$$h_{ib,1} = h_{ib,2} = \frac{0.026}{I_E} = \frac{0.026}{1.405 \text{ mA}} = 18.5 \ \Omega \qquad (\text{A1-11})$$

$$A_{v3} = \frac{e_o}{e_2} \approx \frac{R_p}{h_{ib,2}} = \frac{2423}{18.5} = 130.97 \qquad (\text{A1-1})$$

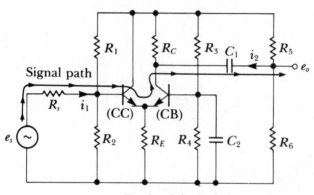

Fig. 1-2   Signal path for Prob. 1.

$$A_{v2} = \frac{e_2}{e_1} \approx \frac{R_{E1}}{h_{ib,1} + R_{E1}} \qquad (\text{A3-1})$$

where $\qquad R_{E1} = R_E \| h_{ib,2} = 1 \text{ k}\Omega \| 18.5 \ \Omega = 18.2 \ \Omega$

$$A_{v2} = \frac{18.2}{18.5 + 18.2} = 0.496$$

[Usually in a differential-pair front end, the gain of the first stage (Q_1) is approximately $\frac{1}{2}$.]

$$A_{v1} = \frac{e_1}{e_s} = \frac{Z_{in\,1} \| R_{12}}{R_s + Z_{in\,1} \| R_{12}}$$

$$Z_{in\,1} \approx -h_{fc,1}(h_{ib,1} + R_{E1}) \qquad (\text{A3-2})$$

Since $h_{fc,1} \approx -h_{fe,1}$,   (A7-23)

$$Z_{in1} = -(-50)(18.5 + 18.2) = 1835 \ \Omega$$

$$R_{12} = 10 \ \text{k}\Omega \| 47 \ \text{k}\Omega = 8246 \ \Omega$$

$$A_{v1} = \frac{8245\|1835}{1000 + 8245\|1835} = 0.6$$

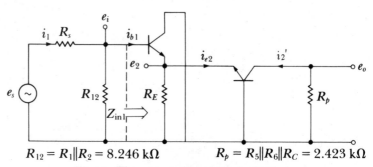

$$R_{12} = R_1\|R_2 = 8.246 \ \text{k}\Omega \qquad R_p = R_5\|R_6\|R_C = 2.423 \ \text{k}\Omega$$

Fig. 1-3   The ac equivalent circuit for Prob. 1.

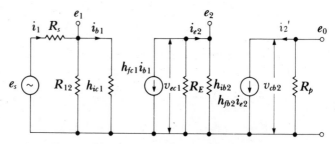

Fig. 1-4   Linear ac equivalent circuit for Prob. 1.

Now the total gain is equal to the product of all the gains:

$$A_{vt} = A_{v1} \times A_{v2} \times A_{v3} = (0.6)(0.496)(130.97)$$
$$= 38.98$$

c. Since the voltage gain is known, the current gain can be determined from Fig. 1-1.

$$A_{vt} = \frac{e_o}{e_s} = 38.98$$

$$e_o = i_2(R_5\|R_6)$$

$$e_s = i_1(R_s + Z_{in1}\|R_{12})$$

$$\frac{e_o}{e_s} = \frac{i_2(R_5\|R_6)}{i_1(R_s + Z_{in1}\|R_{12})} = 38.98$$

$$\frac{i_2}{i_1} = \frac{(2501)(38.98)}{5000} = 19.5$$

CHECK THE SOLUTION: Insert 1 mV into the input, and the following can be solved for: $e_1 = 0.6$ mV; $e_2 = 0.298$ mV; $e_o = 38.98$ mV ($A_{vt} = 38.98$).

$$i_2 = \frac{e_o}{R_5\|R_6} = 7.796 \ \mu\text{A}$$

$$i_1 = \frac{e_s - e_1}{R_s} = 0.4 \times 10^{-6} \quad \text{and} \quad \frac{i_2}{i_1} = 19.49$$

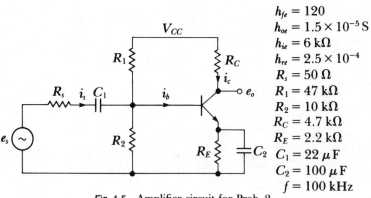

$$h_{fe} = 120$$
$$h_{oe} = 1.5 \times 10^{-5}\,\text{S}$$
$$h_{ie} = 6 \ \text{k}\Omega$$
$$h_{re} = 2.5 \times 10^{-4}$$
$$R_s = 50 \ \Omega$$
$$R_1 = 47 \ \text{k}\Omega$$
$$R_2 = 10 \ \text{k}\Omega$$
$$R_C = 4.7 \ \text{k}\Omega$$
$$R_E = 2.2 \ \text{k}\Omega$$
$$C_1 = 22 \ \mu\text{F}$$
$$C_2 = 100 \ \mu\text{F}$$
$$f = 100 \ \text{kHz}$$

Fig. 1-5   Amplifier circuit for Prob. 2.

PROBLEM 2

GIVEN: The circuit shown in Fig. 1-5.

FIND:

a. Voltage gain $A_v = e_o/e_s$.
b. Current gain $A_i = i_c/i_s$.

SOLUTION: Check to see that C_1 and C_2 are short circuits at 100 kHz. Next, draw the ac equivalent circuit by shorting C_1 and C_2 and making the power supply ac ground.

$$X_{C1} = \frac{1}{2\pi f C_1} = \frac{1}{2\pi(10^5)(22)(10^{-6})} = 0.0723 \ \Omega$$

$$X_{C2} = 0.0159 \ \Omega \qquad \text{similarly}$$

Therefore C_1 and C_2 are negligible at 100 kHz. The ac equivalent circuit is shown in Fig. 1-6, while the linear ac equivalent circuit is shown in Fig. 1-7. This is a common-emitter (CE) amplifier as described in Appendix A2.

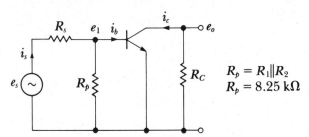

Fig. 1-6   The ac equivalent circuit for Prob. 2.

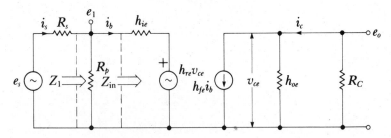

Fig. 1-7   Linear ac equivalent circuit for Prob. 2.

Designate the base as e_1 and solve for the gain expressions as follows:

$$A_v = (A_{v1})(A_{v2})$$

where $\qquad A_{v2} = \dfrac{e_o}{e_1} \qquad$ and $\qquad A_{v1} = \dfrac{e_1}{e_s}$

and $$A_i = (A_{i1})(A_{i2})$$

where $$A_{i1} = \frac{i_b}{i_s} \quad \text{and} \quad A_{i2} = \frac{i_c}{i_b}$$

$$\Delta h_e = h_{ie}h_{oe} - h_{fe}h_{re} \tag{A2-10}$$

$$\Delta h_e = (6 \times 10^3)(1.5 \times 10^{-5}) - (120)(2.5 \times 10^{-4}) = 0.06$$

$$A_{v2} = \frac{-h_{fe}R_C}{h_{ie} + R_C\Delta h_e} = \frac{(-120)(4.7 \times 10^3)}{6 \times 10^3 + (4.7 \times 10^3)(0.06)} = -89.78 \tag{A2-5}$$

$$A_{i2} = \frac{h_{fe}}{1 + R_C h_{oe}} = \frac{120}{1 + (4.7 \times 10^3)(1.5 \times 10^{-5})} = 112.097 \tag{A2-7}$$

$$Z_{in} = \frac{h_{ie} + R_C\Delta h_e}{1 + R_C h_{oe}} = \frac{6 \times 10^3 + (4.7 \times 10^3)(0.06)}{1 + (4.7 \times 10^3)(1.5 \times 10^{-5})} = 5.868 \text{ k}\Omega \tag{A2-6}$$

$$Z_1 = R_1\|R_2\|Z_{in} = 3.428 \text{ k}\Omega$$

$$A_{v1} = \frac{Z_1}{R_S + Z_1} = \frac{3.428 \text{ k}\Omega}{50 + 3.428 \text{ k}\Omega} = 0.9856$$

Since $i_b/i_s = A_{i1}$ and $i_b Z_{in}/[i_s(R_s + Z_1)] = A_{v1}$

$$A_{i1} = A_{v1}\left(\frac{R_S + Z_1}{Z_{in}}\right) = \frac{0.9856(3.478 \times 10^3)}{5.868 \times 10^3} = 0.5842$$

$$A_v = (A_{v1})(A_{v2}) = (0.9856)(-89.78) = -88.487$$

$$A_i = (A_{i1})(A_{i2}) = (0.5842)(112.097) = 65.487$$

CHECK THE SOLUTION: Assume 1 mV input at e_s. $e_1 = A_{v1}e_s = 0.986$ mV; $e_o = A_{v2}e_1 = -88.5$ mV;

$$i_b = \frac{e_1}{Z_{in}} = 0.168 \ \mu A \qquad h_{fe}i_b = 20.16 \ \mu A$$

also,

$$h_{fe}i_b = i_c \times v_{ce}h_{oe} = 20.16 \ \mu A$$

$$Ai_2 = \frac{i_c}{i_b} = 112.08$$

$$i_s = \frac{e_s - e_1}{R_s} = 0.288 \ \mu A$$

and

$$i_s = i_b + \frac{e_1}{R_p} = 0.287 \ \mu A$$

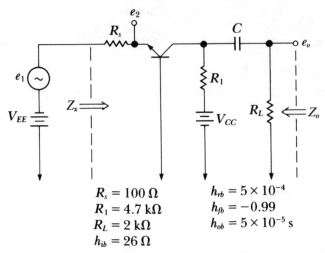

$$R_s = 100 \, \Omega$$
$$R_1 = 4.7 \, \text{k}\Omega$$
$$R_L = 2 \, \text{k}\Omega$$
$$h_{ib} = 26 \, \Omega$$

$$h_{rb} = 5 \times 10^{-4}$$
$$h_{fb} = -0.99$$
$$h_{ob} = 5 \times 10^{-5} \, \text{s}$$

Fig. 1-8 Circuit configuration for Prob. 3.

PROBLEM 3

GIVEN: The circuit of Fig. 1-8 with the values shown. Assume X_C is negligible and neglect any stray capacitances.

FIND:

a. Identify the circuit type.

b. Compute the voltage gain e_o/e_1.

c. Compute Z_o as seen at the output terminals.

d. Compute the input impedance as seen by the signal generator (Z_x).

SOLUTION:

a. This is a common base, NPN transistor amplifier. The ac equivalent circuit is shown in Fig. 1-9 while the linear ac equivalent circuit is shown in Fig. 1-10. The equations are given in Appendix A1.

b. The voltage gain is

$$A_v = \left(\frac{e_o}{e_2}\right)\left(\frac{e_2}{e_1}\right)$$

$$\Delta h_b = h_{ib}h_{ob} - h_{fb}h_{rb} \tag{A1-10}$$

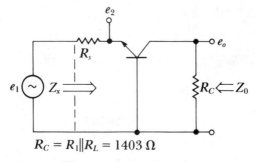

$$R_C = R_1 \| R_L = 1403 \ \Omega$$

Fig. 1-9   The ac equivalent circuit for Prob. 3.

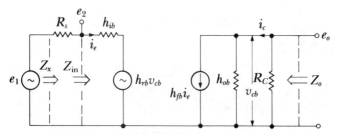

Fig. 1-10   Linear ac equivalent circuit for Prob. 3.

$$\Delta h_b = (26)(5 \times 10^{-5}) - (-0.99)(5 \times 10^{-4})$$

$$\Delta h_b = 1.795 \times 10^{-3}$$

$$\frac{e_o}{e_2} = \frac{-h_{fb}R_C}{h_{ib} + R_C\Delta h_b} = \frac{-(-0.99)(1403)}{26 + (1403)(1.795 \times 10^{-3})} = 48.7 \qquad \text{(A1-5)}$$

$$\frac{e_2}{e_1} = \frac{Z_{in}}{R_s + Z_{in}}$$

$$Z_{in} = \frac{h_{ib} + R_C\Delta h_b}{1 + R_C h_{ob}} = \frac{26 + (1403)(1.795 \times 10^{-3})}{1 + (1403)(5 \times 10^{-5})} = 26.65 \ \Omega \quad \text{(A1-6)}$$

$$\frac{e_2}{e_1} = \frac{26.65}{100 + 26.65} = 0.21$$

$$A_v = \left(\frac{e_o}{e_2}\right)\left(\frac{e_2}{e_1}\right) = 48.7 \times 0.21 = 10.23$$

c. $Z_{ot} = \dfrac{h_{ib} + R_g}{\Delta h_b + R_g h_{ob}} \bigg\| R_C$ (A1-9)

$$= \dfrac{26 + 100}{1.795 \times 10^{-3} + (100)(5 \times 10^{-5})} \bigg\| 1403 = (18.54 \times 10^3) \| 1403$$

$$= 1.304 \text{ k}\Omega$$

Note: $R_s = R_g$.

d. $Z_x = R_s + Z_{in} = 100 + 26.65 = 126.65 \ \Omega$

CHECK THE SOLUTION: Assume e_1 is 1 mV and find e_2 and e_o

$$e_2 = Ae_1 = 0.21 \text{ mV} \qquad e_o = Ae_2 = 10.23 \text{ mV}$$

$$i_e = \dfrac{e_1 - e_2}{R_s} = 7.9 \ \mu\text{A}$$

also,

$$i_e = \dfrac{e_2 - h_{rb}V_{cb}}{h_{ib}} \approx 7.88 \ \mu\text{A}$$

Check the output circuit:

$$\dfrac{-e_o}{R_c} = e_o h_{ob} + h_{fb} i_e = -7.29 \ \mu\text{A vs. } 7.31 \ \mu\text{A}$$

PROBLEM 4

GIVEN: The gain-control circuit of Fig. 1-11. Q_2 is the variable-resistance element and Q_1 and Q_2 are matched. V_1 is the controlling voltage.

FIND:

a. What can be said of the magnitude of V_1, V_2, and e_{in} with respect to distortion?

b. How does the circuit work?

c. What is the value of R_1 for a gain-control range of 4:1 and V_1 variable from -2.5 V to -10 V?

d. Can you recommend two other uses of this circuit?

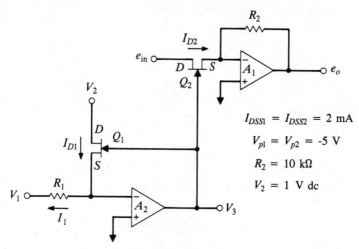

Fig. 1-11 Gain control circuit for Prob. 4.

SOLUTION:

a. V_{DS} must be small with respect to V_p to minimize distortion. Since $V_p = -5$ V, let's keep V_1, V_2, and E_{in} to less than 1 V in magnitude.

b. The current through Q_1 and V_1 establishes a resistance in Q_1 and is shown in Fig. 1-12. The gate-to-source voltage (V_3) adjusts to whatever value is necessary to create this value of resistance. This same V_{GS} is applied to Q_2. Once the gate-to-source voltage has been set, the resistance is independent of applied V_{DS} for small signals.

c. For a gain-control range of $4:1$, r_{DS2} of Q_2 must vary from a minimum of 2.5 kΩ: ($A_v = e_o/e_{in}$)

$$A_v'' = \frac{-R_2}{r_{DS2}'} = -4 \qquad (\text{'' means maximum; ' means minimum})$$

$$r_{DS2}' = r_{DS1}' = \frac{10 \text{ k}\Omega}{4} = 2.5 \text{ k}\Omega$$

to a maximum of 10 kΩ:

$$A_v' = \frac{-R_2}{r_{DS2}''} = -1$$

$$r_{DS2}'' = -R_2/A_v' = \frac{-(10 \text{ k}\Omega)}{-1} = 10 \text{ k}\Omega$$

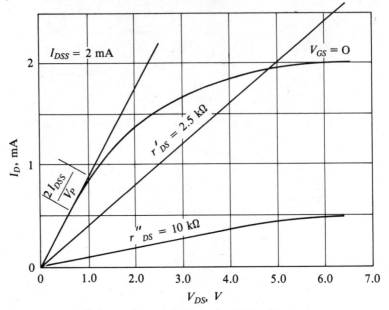

Fig. 1-12   Drain characteristics for JFET of Prob. 4.

Since

$$V_2 = 1 \text{ V} \qquad (\text{Note } r_{DS2} = r_{DS1})$$

$$I''_{D2} = \frac{V_2}{r'_{DS2}} = \frac{1}{2.5 \text{ k}\Omega} = 0.4 \text{ mA}$$

$$I'_{D2} = \frac{V_2}{r''_{DS2}} = \frac{1}{10 \text{ k}\Omega} = 0.1 \text{ mA}$$

Therefore, for $V''_1 = -10$ V:

$$R_1 = \frac{-V_1}{I_{D2}} = \frac{10}{0.4 \text{ mA}} = 25 \text{ k}\Omega$$

and

$$V_1 = 2.5 \text{ V} \qquad \text{for } I_{D2} = 0.1 \text{ mA}$$

d.   It can be used as an analog multiplier or divider.

$$I_1 = I_{D1} = \frac{0 - V_1}{R_1} = \frac{-V_1}{R_1} \qquad (V_1 \text{ is negative})$$

$$r_{DS1} = \frac{V_2}{I_{D1}} = \frac{-V_2}{V_1} R_1 = r_{DS2}$$

$$e_{in}/r_{DS2} = I_{D2} \quad \text{and} \quad -I_D R_2 = e_o$$

$$e_o = \frac{-e_{in}R_2}{r_{DS2}} = \frac{e_{in}V_1R_2}{V_2R_1} = \frac{Ke_{in}V_1}{V_2}$$

CHECK THE SOLUTION: Check each step numerically. Also the minimum achievable value of r_{DS} for the I_{DSS} and V_p given is approximately $r_{DS} = |V_p|/I_{DSS} = 5/(0.002) = 2.5$ kΩ which is acceptable since the minimum value of r_{DS} that we need is 2.5 kΩ. The slope at $V_{DS} = 0$ is actually $|V_p|/2I_{DSS}$, but it's safer to use $|V_p|/I_{DSS}$ for larger signal swings.

PROBLEM 5

GIVEN: The circuit of Fig. 1-13. This circuit can be used to synthesize an inductor.

FIND: The value of C to yield an inductor of 100 mH.

SOLUTION: Let's assume an ideal amplifier and label the nodes as shown in Fig. 1-14. Remember that the voltage at the $(+)$ input of the amplifier is equal in magnitude to that at the $(-)$ input for proper

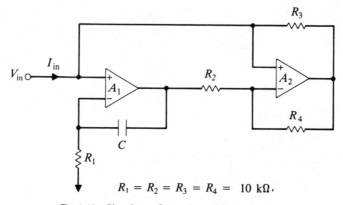

$$R_1 = R_2 = R_3 = R_4 = 10 \text{ k}\Omega.$$

Fig. 1-13   Circuit configuration of Prob. 5.

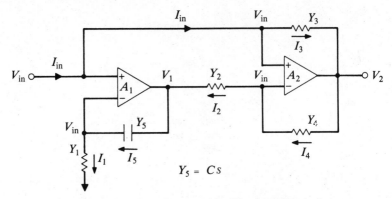

Fig. 1-14   Node and current identification of Prob. 5.

operation, and the input currents are zero. At the $(+)$ input of A_2:

$$(V_{in} - V_2)Y_3 = I_{in}$$
$$V_2Y_3 = V_{in}Y_3 - I_{in}$$
$$V_2 = V_{in} - I_{in}Z_3 \qquad (1)$$

At the $(-)$ input of A_1:

$$(V_{in} - 0)Y_1 = (V_1 - V_{in})Y_5$$
$$V_{in}Y_1 = V_1Y_5 - V_{in}Y_5$$
$$V_1Y_5 = V_{in}(Y_1 + Y_5)$$
$$V_1 = V_{in}(Y_1 + Y_5)Z_5 \qquad (2)$$

At the $(-)$ input of A_2:

$$(V_{in} - V_1)Y_2 = (V_2 - V_{in})Y_4$$
$$V_{in}Y_2 - V_1Y_2 = V_2Y_4 - V_{in}Y_4$$
$$V_{in}(Y_2 + Y_4) - V_1Y_2 - V_2Y_4 = 0 \qquad (3)$$

(1) and (2) into (3):

$$V_{in}(Y_2 + Y_4) - V_{in}(Y_1 + Y_5)Z_5Y_2 - (V_{in} - I_{in}Z_3)Y_4 = 0$$
$$V_{in}(Y_2 + Y_4 - Y_1Y_2Z_5 - Y_2Y_5Z_5 - Y_4) + I_{in}Z_3Y_4 = 0$$
$$V_{in}(Y_2 - Y_1Y_2Z_5 - Y_2) = -I_{in}Z_3Y_4$$

and

$$\frac{V_{in}}{I_{in}} = \frac{Z_3 Y_4}{Y_1 Y_2 Z_5} = \frac{Y_4 Y_5}{Y_1 Y_2 Y_3}$$

Substituting $R_1 = R_2 = R_4 = 10 \text{ k}\Omega$, and $Y_5 = Cs$:

$$\frac{V_{in}}{I_{in}} = \frac{10^{-4} Cs}{10^{-12}} = 10^8 Cs$$

For this problem, $L = 0.1$ H

$$0.1s = 10^8 Cs$$

$$C = \frac{0.1}{10^8} = 10^{-9} \text{ F}$$

CHECK THE SOLUTION: Check at a convenient frequency where $1/\omega C = 10 \text{ k}\Omega$: choose $\omega = 10^5$ rad/s. If $V_{in} = 1$ V, $X_c = -j10^4$ and $I_1 = I_s = 10^{-4}$; $V_1 = 1 - j$; $I_2 = I_4 = j10^{-4}$; $V_2 = 1 + j1$; $I_{in} = -j10^{-4}$; and $Z_{in} = V_{in}/I_{in} = j10^4 = j\omega L$; and $L = 0.1$ H.

PROBLEM 6

GIVEN: The circuit and amplifier characteristics of Fig. 1-15. Assume the effects due to gain, output impedance, and input impedance are negligible.

FIND:

a. The nominal output voltage.

b. The change in output due to the change in zener diode voltage, offset current, bias current, and power-supply rejection; each computed separately.

SOLUTION:

a. The nominal output voltage can be determined by comparing this circuit with the basic op-amp circuit of Fig. C4-1:

$$E_o = \frac{(V_1 - V_2)(R_3 + R_4)R_2}{(R_1 + R_2)R_3} + \frac{(R_3 + R_4)V_2}{R_3} - \frac{V_3 R_4}{R_3} \quad \text{(C3-1)}$$

For the circuit of Fig. C3-1 of Appendix C3, and Eq. (C3-1)

$$V_1 = V_Z \qquad V_2 = V_3 = 0 \text{ V dc}$$

$$R_1 = 0 \qquad R_2 = \infty \qquad R_3 = 10 \text{ k}\Omega \qquad R_4 = 4.7 \text{ k}\Omega$$

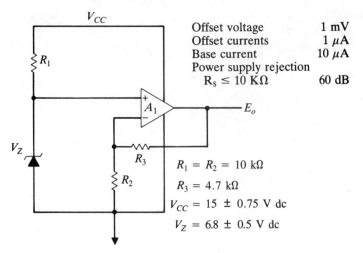

Offset voltage — 1 mV
Offset currents — 1 μA
Base current — 10 μA
Power supply rejection
$R_S \le 10 \text{ K}\Omega$ — 60 dB

$R_1 = R_2 = 10 \text{ k}\Omega$

$R_3 = 4.7 \text{ k}\Omega$

$V_{CC} = 15 \pm 0.75 \text{ V dc}$

$V_Z = 6.8 \pm 0.5 \text{ V dc}$

Fig. 1-15 Circuit configuration for Prob. 6.

Therefore:

$$E_0 = \frac{V_1(R_3 + R_4)}{R_3} = \frac{6.8(14.7 \text{ k}\Omega)}{10 \text{ k}\Omega} = 9.996 \text{ V dc}$$

b. The total effect due to all our sources except power-supply rejection is given by

$$V_{ost} = \left[V_{os} + I_B(R_S - R_{eq}) + \frac{I_{os}(R_S + R_{eq})}{2} \right] \frac{R_f + R_i}{R_i} \quad \text{(C4-1)}$$

where $R_S = R_z = 0$ and $R_{eq} = R_3 \| R_4 = 3.2 \text{ k}\Omega$

The output change due to the offset voltage is

$$V_{ost1} = \frac{V_{os}(R_f + R_i)}{R_i} = \frac{0.005 \text{ V}(14.7 \text{ k}\Omega)}{10 \text{ k}\Omega} = 7.35 \text{ mV}$$

The output offset voltage due to the base current is

$$V_{ost2} = \frac{I_B(R_S - R_{eq})(R_f + R_i)}{R_i}$$

$$= \frac{(-10^{-5})(-3.2 \text{ k}\Omega)(14.7 \text{ k}\Omega)}{10 \text{ k}\Omega} = 47 \text{ mV}$$

The output offset voltage due to the offset current:

$$V_{ost3} = \frac{I_{os}(R_S + R_{eq})(R_f + R_i)}{2R_i} = \frac{10^{-6}(3.2 \text{ k}\Omega)(14.7 \text{ k}\Omega)}{20 \text{ k}\Omega} = 2.35 \text{ mV}$$

The output offset due to the power-supply rejection can be determined as follows: If we have 60 dB of power-supply rejection, then for every volt change of V_{cc} we get 1 mV change at the input terminals (60 dB = 1000:1; 1 V/1000 = 1 mV). If we have a 0.75-V change in V_{cc} we obtain a change in the output of:

$$\Delta V_o = \frac{(0.75 \text{ V})(R_f + R_i)}{1000 \ R_i} = \frac{(0.75)(14.7 \text{ k}\Omega)}{10 \text{ M}\Omega} = 1.1 \text{ mV}$$

CHECK THE SOLUTION: If E_o is 9.996, then I_{R3} can be calculated:

$$I_{R3} = \frac{E_o - V_Z}{R_3} = \frac{(9.996 - 6.8) \text{ V}}{4.7 \text{ k}\Omega} = 0.68 \text{ mA}$$

I_{R3} must equal I_{R2} since there is no current into the negative terminal.

$$I_{R2} = \frac{V_Z}{R_2} = \frac{6.8 \text{ V}}{10 \text{ k}\Omega} = 0.68 \text{ mA}$$

The remainder of the problem can be checked by checking each solution numerically.

PROBLEM 7

GIVEN: The circuit of Fig. 1-16. The diode is placed across the solenoid to protect the transistor during turn-off. The relay coil resistance is 30 Ω and its inductance is 10 mH. The breakdown of the transistor is 50 V (BV_{CEO}). V_{CC} = 20 V dc.

FIND: Will the solenoid deactivate if the turn-off pulse is applied for 0.5 ms? Assume the drop-out current is 100 mA. If not, what can we do to increase the turn-off speed? Assume the forward drop of CR_1 (V_{CR1}) is 1 V.

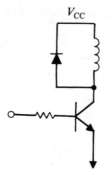

Fig. 1-16   Circuit of Prob. 7.

SOLUTION: At turn-off, the equivalent curcuit is shown in Fig. 1-17. We can write the defining equation as follows (assume $V_{CE\,sat} = 0.4$ V dc):

$$V(s) = L[sI(s) - i(0^+)] + RI(s) \qquad (1)$$

where

$$V(s) = \frac{V_{CC} + V_{CR1} - V_{CC}}{s} = \frac{V_{CR1}}{s} = \frac{1}{s} \qquad (2)$$

and

$$i(0^+) = \frac{-(V_{CC} - V_{CEsat})}{R} = \frac{-(20 - 0.4)}{30} = -0.6533 \text{ A} \qquad (3)$$

Combine (1), (2), (3)

$$1/s = 0.01[sI(s) + 0.6533] + 30I(s)$$

$$1/s = (0.01s + 30)I(s) + 6.533 \times 10^{-3}$$

$$1 - 6.533 \times 10^{-3}s = 0.01s(s + 3000)I(s)$$

$$I(s) = \frac{1}{0.01s(s + 3000)} - \frac{6.533 \times 10^{-3}}{0.01(s + 3000)}$$

$$i(t) = \frac{1}{30} - 0.6866e^{-3000t}$$

Make two cursory checks; at $t = 0$,

$$i(o) = \frac{1}{30} - 0.6866 = -0.6533 \text{ A}$$

at $t = \infty$,

$$i(\infty) = \frac{1}{30} = 0.0333 \text{ A}$$

and both of these check. At $i(t) = 100$ mA, let's solve for t. Note $i(t) = -100$ mA because current flow opposes the source of V_{CR1}.

$$-0.1 = \frac{1}{30} - 0.6866e^{-3000t}$$

$$-0.133 = -0.6866e^{-3000t}$$

$$t = \frac{\ln(0.1937)}{-3000} = \frac{-1.64}{-3000}$$

$$t = 0.547 \text{ ms}$$

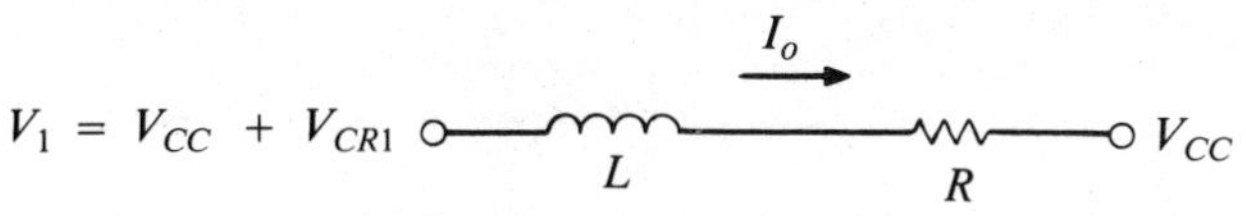

$$V_1 = V_{CC} + V_{CR1} \quad \text{o————} L \text{————} R \text{————o } V_{CC}$$

$$I_o \longrightarrow$$

Fig. 1-17 Equivalent circuit for Fig. 1-16.

And we will not turn off during the specified time of 0.5 ms. We can increase the speed by using any of the circuits a, b, or c of Fig. 1-18 as possible candidates. VR_1 or V_{clamp} must be selected to keep the collector voltage of the transistor to less than its breakdown BV_{CEO}. The analysis for all circuits will be similar, but let's choose the circuit of Fig. 1-18a. The equivalent circuit is shown in Fig. 1-19. Select the zener voltage V_z to be 15 V, substitute into our earlier equation, and solve for t.

$$i(t) = \frac{V_z}{30} - \left[\frac{V_z}{R} + i(0^+) \right] e^{-\frac{Rt}{L}}$$

$$-0.1 = \frac{15}{30} - \left(\frac{15}{30} + 0.6533 \right) e^{-3000t}$$

$$-0.6 = -1.1533 e^{-3000t}$$

$$t = \frac{\ln(0.52)}{-3000} = 0.218 \text{ ms}$$

which meets our 0.5 ms turn-off criteria. (Note: In actual practice just meeting the BV_{CEO} parameter is not sufficient. We should make sure we do not create a secondary breakdown condition. Normally, SOAR curves

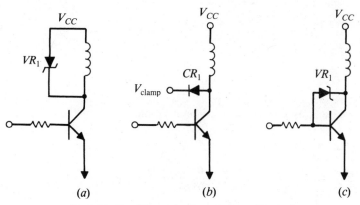

Fig. 1-18 Candidates for the solution of Prob. 7.

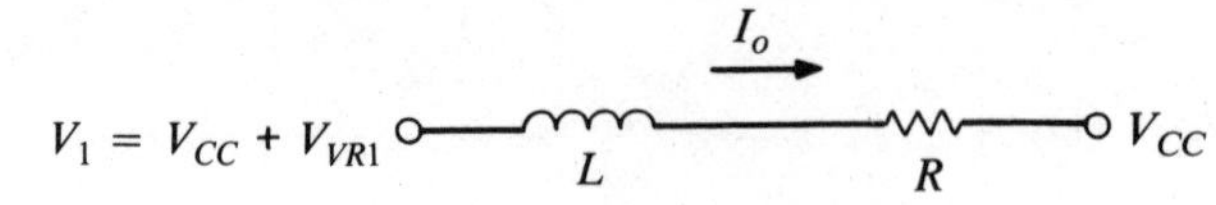

$$V_1 = V_{CC} + V_{VR1}$$

Fig. 1-19   Equivalent circuit for Fig. 1-18a.

can be obtained that indicate the amount of inductive switching a transistor can handle.)

CHECK THE SOLUTION: At $t = 0.218$ ms

$$i(t) = \frac{15}{30} - \left(\frac{15}{30} + 0.6533\right) e^{-3000(0.218 \times 10^{-3})}$$

$$i(t) - 0.5 - 0.6 = -0.1 \text{ A}$$

and this checks our solution.

PROBLEM 8

GIVEN: The buffer amplifier shown in Fig. 1-20. The minimum input signal is -5 V dc and the maximum is 0 V dc.

FIND: If the JFETs are perfectly matched with the parameters shown, what is the maximum output error contributed by the JFETs?

SOLUTION: Since the JFETs are perfectly matched, $V_{GS1} = V_{GS2}$ as long as the drain currents are the same, and the output voltage will be equal to the input voltage. When the drain currents become mismatched, an offset error is introduced because $V_{GS1} \neq V_{GS2}$. When $e_o = 0$ V dc, $I_{D1} = I_{D2}$, and $V_{GS1} = V_{GS2} = 0$. The worst-case error is introduced when e_o is maximum negative. The current through R_L is

$$I_{RL} = \frac{5}{100 \times 10^3} = 0.05 \text{ mA}$$

$$I_{D2} = I_{DSS2} = 1 \text{ mA}$$

$$I_{D1} = I_{D2} - I_{RL} = 1 - 0.05 = 0.95 \text{ mA}$$

We need to find the change in V_{GS} for a change in I_D. This can be done by finding the g_{fs} at the operating point of $I_D = I_{DSS}$.

$$g_{fso1} = \frac{\Delta I_{D1}}{\Delta V_{GS1}} = \frac{2 I_{DSS1}}{|V_{p1}|} = \frac{2 \text{ mA}}{3 \text{ V}} = 667 \ \mu\text{S*} \qquad \text{(F1-3)}$$

$$\Delta V_{GS1} = \frac{\Delta I_{D1}}{g_{fso1}} = \frac{-50 \times 10^{-6}}{667 \times 10^{-6}} = -0.075 \text{ V dc}$$

*S is the abbreviation for siemens, the SI unit of conductance. It is equivalent to mho.

This is our maximum output error. Our absolute output voltage is

$$e_o = e_{in} - V_{GS1} = -5 - (-0.075) = -4.925 \text{ V dc}$$

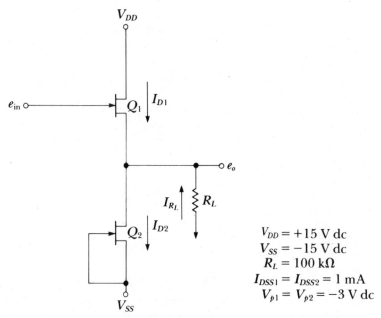

$$V_{DD} = +15 \text{ V dc}$$
$$V_{SS} = -15 \text{ V dc}$$
$$R_L = 100 \text{ k}\Omega$$
$$I_{DSS1} = I_{DSS2} = 1 \text{ mA}$$
$$V_{p1} = V_{p2} = -3 \text{ V dc}$$

Fig. 1-20   Buffer amplifier for Prob. 8.

CHECK THE SOLUTION: The drain current can be checked very easily by using Eq. (F1-1):

$$I_{D1} = I_{DSS1} \left(1 - \frac{V_{GS}}{V_p} \right)^2 \tag{F1-1}$$

$$I_{D1} = 1 \text{ mA} \left(1 - \frac{-0.075}{-3} \right)^2 = (1 \text{ mA})(0.951) = 0.9506 \text{ mA}$$

which is a close check. The small amount of error is attributable to the fact that g_{fs} varies with I_D, and the further we move from the operating point the greater the error.

PROBLEM 9

GIVEN: The gain of the transistor amplifier of Fig. 1-21 is to be varied by controlling the dynamic resistance of the forward-biased diode, CR_1.

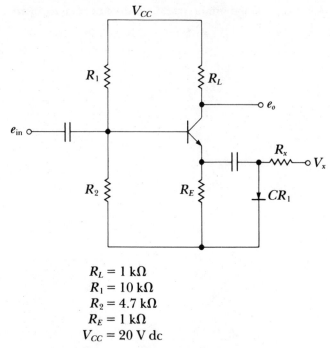

$$R_L = 1 \text{ k}\Omega$$
$$R_1 = 10 \text{ k}\Omega$$
$$R_2 = 4.7 \text{ k}\Omega$$
$$R_E = 1 \text{ k}\Omega$$
$$V_{CC} = 20 \text{ V dc}$$

Fig. 1-21   Gain-controlled amplifier of Prob. 9.

FIND: If we want a gain change from $A_v = 1$ to 10, what value must R_x' be for a control voltage range of $V_x = 0$ to 10 V dc?

SOLUTION: Assume the reactance of the capacitor to be $0 \, \Omega$. From Appendix A5, the approximate gain of a CE amplifier is:

$$A_v \approx -R_C/(h_{ib} + R_E) \qquad \text{(A5-1)}$$

h_{ib} is a function of I_E. Let's use approximate analysis to determine I_E:

$$V_B \approx (R_2)\,(V_{CC})/(R_1 + R_2) = 6.39 \text{ V dc}$$
$$V_E \approx V_B - V_{BE} = 6.39 - 0.7 = 5.69 \text{ V dc}$$
$$I_E = 5.69 \text{ V}/(1 \text{ k}\Omega) = 5.69 \text{ mA}$$
$$h_{ib} = 0.026/(5.69 \text{ mA}) = 4.57 \, \Omega$$

Let $R_E' = R_E \| r_d$, where r_d is the dynamic impedance of the diode and is equal to:

$$r_d = 0.026/I_D$$

I_D is the direct current flowing through the diode. For a gain of 1,

$$1 = R_C/(h_{ib} + R'_{E(max)})$$
$$R'_{E(max)} = 1\ k\Omega - 4.57 = 995.4\ \Omega$$
$$r_{d(max)} \| R_E = 995.4\ \Omega$$

Solving, we find

$$r_{d(max)} = 216.4\ k\Omega$$
$$I_{D(min)} = 0.026/r_{d(max)} = 0.026/216.4\ k\Omega = 0.12\ \mu A$$

This is theoretically correct, but we should start investigating leakage currents because of the small value of I_D.

For a gain of 10,

$$10 = R_C/(h_{ib} + R'_{E(min)})$$
$$R'_{E(min)} = 95.43\ \Omega$$
$$r_d \| R_E = 95.43\ \Omega$$
$$r_{d(min)} = 105.5\ \Omega$$
$$I_{D(max)} = 0.026/r_{d(min)} = 0.026/105.5 = 0.246\ mA$$

Determine R_x; assume $V_{CR1} = 0.7\ V$ dc

$$R_x = (V_x - V_{CR1})/0.246\ mA = 37.8\ k\Omega$$

Use the nearest standard value of 37.4 kΩ found in Appendix H1.

CHECK THE SOLUTION: At $I_D = 0.12\ \mu A$,

$$r_{d(max)} = 0.026/(0.12 \times 10^{-6}) = 216.67\ k\Omega$$
$$r_d \| R_E = 216.67\ k\Omega \| 1\ k\Omega = 995.4\ \Omega$$

and this checks. At $I_D = 0.246\ mA$, $r_{d(min)} = 0.026/0.000246 = 105.7\ \Omega$.

$$r_d \| R_E = 105.7\ \Omega \| 1\ k\Omega = 95.6\ \Omega$$

which is a close check.

PROBLEM 10

GIVEN: A sinusoidal voltage source e_i of magnitude 50 V peak and frequency of 1 kHz.

FIND: The output e_o for the three circuits of Fig. 1-22 and sketch the waveforms. Assume the diode forward drop is 1 V, and the generator source resistance R_S is 50 Ω.

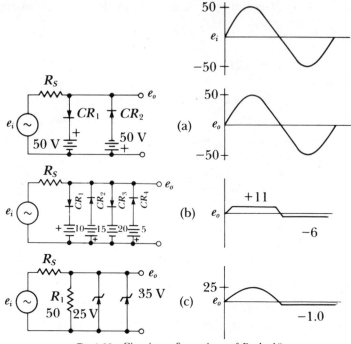

Fig. 1-22   Circuit configurations of Prob. 10.

SOLUTION:

a.   The two diodes CR_1 and CR_2 will clamp the source at $+51$ V and -51 V, respectively when the diode drop is included. Therefore no clamping occurs.

b.   When there are several clamping circuits, the clamp closest to ground will predominate. Therefore the output will be clamped at $+11$ V (CR_1) and -6 V (CR_4).

c.   The resistor limits the swing to $[R_1/(R_s + R_1)]e_i$, which is one-half of V_i. Therefore, neither zener will clamp the input in the positive direction. In the negative direction, the zeners will become forward-biased and will act as a normal diode to clip at -1 V. (Some temperature-compensated diodes look like a zener diode in both directions, but this is the exception rather than the rule.) If our signal were to have been $+100$ V in, the 25-V zener would have clamped in the positive direction, because it would have started conducting before the 35-V zener.

PROBLEM 11

GIVEN: The amplifier of Fig. 1-23.

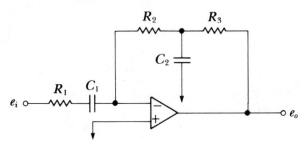

Fig. 1-23   Amplifier circuit of Prob. 11.

FIND: The gain expression e_o/e_i.

SOLUTION: Assume an ideal operational amplifier. Let

$$Z_1 = R_1 + \frac{1}{C_1 s} = \frac{R_1 C_1 s + 1}{C_1 s}$$

Let's define Z_2 as the equivalent feedback impedance of the feedback network. A delta-Y transformation described in Appendix D1 will simplify calculations considerably. Converting the Y feedback network to a delta yields the circuit of Fig. 1-24.

Z_A and Z_C do not contribute to the output voltage because Z_A has a source voltage of 0 V, and Z_C is connected across the output. Since the

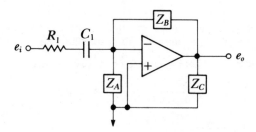

Fig. 1-24   Δ Equivalent circuit of Prob. 11.

output impedance is equal to zero, there is no signal degradation due to Z_C.

$$Z_B = \frac{Z_1 Z_2 + Z_2 Z_3 + Z_1 Z_3}{Z_3} \qquad \text{(D1-5)}$$

$$= \frac{R_2 R_3 + R_3(1/C_2 s) + R_2(1/C_2 s)}{1/C_2 s}$$

$$= R_2 R_3 C_2 s + R_3 + R_2$$

$$\frac{e_o}{e_i} = -\frac{Z_B}{Z_1} = \frac{-(R_2 R_3 C_2 s + R_2 + R_3)C_1 s}{(R_1 C_1 s + 1)} \qquad \text{(C2-1)}$$

CHECK THE SOLUTION: Let's check using Appendix J. Appendix Eq. (J1-3) describes Z_{T2} as

$$Z_{T2} = A(1 + sT) = 2R \left(1 + \frac{sRC}{2}\right)$$

$$Z_{T2} = 2R + R^2 Cs$$

which is of the form of Z_B, which checks Z_B.

PROBLEM 12

GIVEN: Two JFETs with the following parameters as shown in Fig. 1-25.

$$V_{p1} = -4\text{ V} \qquad R_D = 1\text{ k}\Omega$$
$$I_{DSS1} = 3\text{ mA} \qquad R_S = 4.7\text{ k}\Omega$$
$$V_{p2} = -5\text{ V} \qquad V_{SS} = -5\text{ V dc}$$
$$I_{DSS2} = 4\text{ mA} \qquad V_{DD} = +20\text{ V dc}$$

FIND: Which position should the two JFETs be in for maximum gain? Assume $X_C = 0$.

SOLUTION: First identify the two stages and draw the ac equivalent circuit. We have a common-source (CS) stage driving a common-gate (CG) stage. The ac equivalent circuit is shown in Fig. 1-26.

From Appendix B (approximate equations will do):

$$A_{vB} = \frac{e_o}{e_a} \approx \frac{g_{fsB} R_D}{1 + g_{fsB} R_X} \approx g_{fsB} R_D \qquad R_X = R_S = 0\ \Omega \qquad \text{(B2-5)}$$

$$Z_{inB} = R_X + \frac{1}{g_{fsB}} \approx \frac{1}{g_{fsB}} \qquad \text{(B2-6)}$$

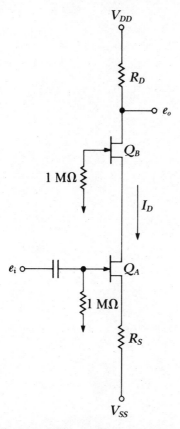

Fig. 1-25   JFET amplifier of Prob. 12.

For the gain equation of Q_A, refer to Appendix B1:

$$A_{vA} = \frac{e_o}{e_i} = -\frac{g_{fsA}R_D}{1 + g_{fsA}R_S} \qquad \text{(B1-5)}$$

But R_D of Eq. (B1-5) is the input impedance of Q_B, which is Z_{inB}:

$$A_{vA} = -\frac{g_{fsA}(1/g_{fsB})}{1 + g_{fsA}R_S}$$

$$A_{vA} = \frac{-g_{fsA}}{g_{fsB} + g_{fsA}g_{fsB}R_S}$$

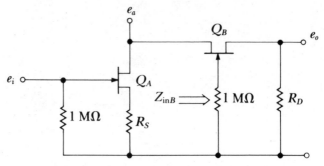

Fig. 1-26   JFET ac equivalent circuit of Prob. 12.

The total gain A_{vt} is

$$A_{vt} = A_{vA} \times A_{vB} = -\frac{g_{fsA}(g_{fsB}R_D)}{g_{fsB} + g_{fsA}g_{fsB}R_S}$$

$$A_{vt} = \frac{-R_D}{(1/g_{fsA}) + R_S}$$

As g_{fsA} becomes very large, A_{vt} approaches $-R_D/R_S$:

$$A_{vt}\Big|_{g_{fsA} \to \infty} = -\frac{R_D}{R_S}$$

Therefore we want g_{fsA} to be larger than g_{fsB}.
    From Appendix F,

$$g_{fso} = \frac{2I_{DSS}}{|V_p|} \tag{F1-3}$$

Since I_{DA} must equal I_{DB}, we can determine which JFET has the best g_{fso}. [If the drain currents were different, we would need to use Eq. (F1-5) at the operating point to determine which has the better g_{fs}.]

$$g_{fso1} = \frac{6\text{ mA}}{4\text{ V}} = 1.5 \times 10^{-3} \tag{F1-5}$$

$$g_{fso2} = \frac{8\text{ mA}}{5\text{ V}} = 1.6 \times 10^{-3}$$

Therefore we place Q_2 in the place of Q_A because it has the better g_{fso}, and consequently the better g_{fs}, since I_D is the same for both.

CHECK THE SOLUTION: Check each step for derivation accuracy.

PROBLEM 13 _____

GIVEN: It is desired to synthesize an ac equivalent circuit for an operational amplifier using a program similar to ECAP. This program has only current generators for its dependent sources. The circuit of Fig. 1-27 may be used to synthesize the operational amplifier.

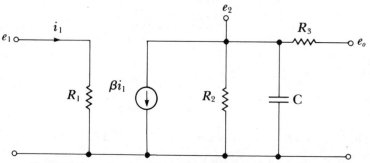

Fig. 1-27   Operational amplifier ac equivalent circuit of Prob. 13.

FIND: All circuit values if the amplifier has the following characteristics:

$$R_{in} = 10^6 \, \Omega \qquad \text{(amplifier input impedance)}$$
$$A_v = 92 \, \text{dB} \qquad \text{(amplifier voltage gain)}$$
$$f_c = 100 \, \text{Hz} \qquad \text{(3-dB pole)}$$
$$r_o = 100 \, \Omega \qquad \text{(output impedance)}$$

SOLUTION: R_1 is equal to the input impedance, therefore

$$R_1 = 10^6 \, \Omega$$

Next select R_3 as the output impedance and select $R_2 \ll R_3$ so that it will cause $(\beta i_1)(R_2)$ to look like a voltage source.

$$R_3 = r_o = 100 \, \Omega$$

and pick $R_2 = 0.001 \, \Omega$.

Find β. Since no current is flowing in the output, $e_o = e_2$. Therefore

$$A_v = \frac{e_o}{e_1} = \frac{e_2}{e_1}$$

$$20 \log A_v = 92 \text{ dB}$$

$$A_v = 10^{92/20}$$

$$= 39{,}811$$

$$e_2 = (\beta i_1)(R_2)$$

$$e_1 = i_1 R_1$$

$$\frac{e_2}{e_1} = \frac{(\beta i_1)R_2}{i_1 R_1} = \frac{\beta R_2}{R_1} = A_v$$

or

$$\beta = \frac{A_v R_1}{R_2}$$

$$= \frac{(39{,}811)(10^6)}{10^{-3}} = 3.9811 \times 10^{13}$$

The breakpoint at 100 Hz is obviously determined by C and R_2.

$$e_2 = (\beta i_1)(Z_2)$$

where

$$Z_2 = \frac{R_2(1/Cs)}{R_2 + 1/Cs} = \frac{1/C}{s + 1/R_2 C}$$

Therefore, at $s = j\omega$

$$\omega = \frac{1}{R_2 C}$$

and

$$C = \frac{1}{\omega R_2} = \frac{1}{2\pi(100)(10^{-3})} = 1.59 \text{ F}$$

CHECK THE SOLUTION:

$$e_2 = (\beta i_1)R_2 = (3.9811 \times 10^{13})(10^{-3})i_1 = (3.9811 \times 10^{10})i_1 \quad \text{(checks)}$$

$$e_1 = i_1 R_1 \quad = (10^6)i_1$$

$$A_v = \frac{e_2}{e_1} \quad = \frac{3.9811 \times 10^{10}}{(10^6)i_1}i_1 \quad = 3.9811 \times 10^4 \quad \text{(checks)}$$

$$X_C = R_2 \quad \text{at } f = 100 \text{ Hz}$$

Since it is the 3-dB point,

$$X_C = \frac{1}{2\pi f C} = \frac{1}{2\pi(100)(1.59)} = 0.001 \qquad \text{(checks)}$$

PROBLEM 14

GIVEN: The amplifier shown in Fig. 1-28. Neglect I_{GSS}.

FIND: The gain e_o/e_i, given the circuit parameters shown.

SOLUTION: R_3 can be neglected because $I_{GSS} = 0$ and the voltage across R_3 is zero. To solve for the gain, we must find the value of g_{fs} which is a function of the dc operating current I_D. We can find I_D from Eqs. (F1-11), (F1-22), and (F1-23) as follows:

$$\alpha = 1 - \frac{R_2 V_1}{V_p(R_1 + R_2)} - \frac{R_1 V_2}{V_p(R_1 + R_2)} + \frac{V_{SS}}{V_p} \qquad \text{(F1-22)}$$

$$= 1 - \frac{(220 \times 10^3)(12)}{-5(1.22 \times 10^6)} - \frac{-5 \times 10^6}{-5(1.22 \times 10^6)} + 0$$

$$= 0.613$$

$$\beta = \frac{R_4}{V_p} = \frac{R_S}{V_p} = \frac{1 \times 10^3}{-5} = -200 \qquad \text{(F1-23)}$$

$$I_D = -\frac{\left(\frac{2\alpha}{\beta} - \frac{1}{\beta^2 I_{DSS}}\right) \pm \left[\left(\frac{2\alpha}{\beta} - \frac{1}{\beta^2 I_{DSS}}\right)^2 - \frac{4\alpha^2}{\beta^2}\right]^{1/2}}{2} \qquad \text{(F1-11)}$$

$$= \frac{(8.63 \pm 6.07)\,\text{mA}}{2}$$

$$= 1.28\,\text{mA},\ 7.35\,\text{mA}$$

Try 1.28 mA as the solution. Check using Eqs. (F1-19), (F1-20), and (F1-1).

$$V_G = \frac{R_2(V_1 - V_2)}{R_1 + R_2} + V_2 \qquad \text{(F1-19)}$$

$$= \frac{0.22 \times 10^6(17)}{1.22 \times 10^6} - 5 = -1.934\,\text{V dc}$$

$$V_S = I_D R_4 + V_{SS} = I_D R_S + V_{SS} \qquad \text{(F1-20)}$$

$$= (1.28 \times 10^{-3})(1 \times 10^3) + 0 = 1.28\,\text{V dc}$$

$$V_{GS} = -1.934 - 1.28 = -3.214$$

Checking with (F-1):

$$I_D = I_{DSS}\left(1 - \frac{V_{GS}}{V_p}\right)^2 = 0.01\left(1 - \frac{-3.214}{-5}\right)^2 \qquad \text{(F1-1)}$$

$$= 1.28\,\text{mA} \qquad \text{(this checks)}$$

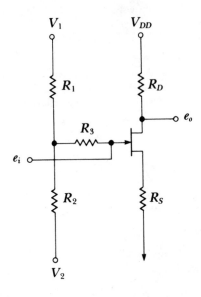

$$
\begin{aligned}
I_{DSS} &= 10\,\text{mA} & R_1 &= 10^6\,\Omega \\
V_{GS(off)} &= -5\,\text{V}\,(V_p) & R_2 &= 220\,\text{k}\Omega \\
V_1 &= +12\,\text{V dc} & R_3 &= 10^6\,\Omega \\
V_{DD} &= +40\,\text{V dc} & R_D &= 10\,\text{k}\Omega \\
V_2 &= -5\,\text{V dc} & R_S &= 1\,\text{k}\Omega
\end{aligned}
$$

Fig. 1-28   JFET amplifier of Prob. 14.

Since we know I_D, we can get g_{fs} as follows:

$$g_{fs} = \frac{2I_{DSS}}{|V_p|}\left(\frac{I_D}{I_{DSS}}\right)^{1/2} \qquad \text{(F1-5)}$$

$$= \frac{2(0.01)}{5}\left(\frac{1.28\,\text{mA}}{10\,\text{mA}}\right)^{1/2} = 1.43 \times 10^{-3}\,\text{A/V}$$

The gain can be determined from Eq. (B1-1) as follows:

$$A_v = \frac{-g_{fs}R_D}{1 + g_0(R_D + R_S) + g_{fs}R_S} \qquad \text{(B1-1)}$$

Since g_o is not given, let's assume its value to be the typical value given in Appendix B1:

$$g_o = 5 \times 10^{-5}$$

$$A_v = \frac{-1.43 \times 10^{-3}(10 \times 10^3)}{1 + 5 \times 10^{-5}(11 \times 10^3) + (1.43 \times 10^{-3})(1 \times 10^3)}$$

$$= \frac{-14.3}{2.98} = -4.8$$

CHECK THE SOLUTION: If we had selected 7.35 mA as a solution, it would not have checked out. In general, the lower value of I_D will be the correct selection. The absolute value of $V_{DS} = (V_D - V_S)$ must be larger in magnitude than the absolute value of pinch-off voltage for the analysis to be valid.

$$V_D = V_{DD} - I_D R_D$$
$$V_D = 40 - (1.28 \text{ mA})(10 \text{ k}\Omega) = 27.2 \text{ V dc}$$
$$V_{DS} = V_D - V_S = 27.2 - 1.28 = 25.92 \text{ V dc}$$

which is much larger than our pinch-off voltage of 5 V. Our check is thus complete.

PROBLEM 15

GIVEN: The amplifier chain shown in Fig. 1-29. All amplifiers are powered by +15 and −15 V and the useful output swing is ±10 V dc.

FIND: Calculate the dc output V_o. What happens if we change R_6 to 40 kΩ?

SOLUTION: The important thing to remember when dealing with dc operation of operational amplifiers is that the forcing voltage is that voltage present at the (+) terminal of the amplifier. Whatever voltage is present on the (+) terminal is present on the (−) terminal. Establish the voltage designations as shown in Fig. 1-30 for ease of description. Assume an ideal operational amplifier as described in Appendix C3.

Define:

$$V_a = V_2 \qquad I_{R1} = \text{current through } R_1$$
$$V_d = V_f \qquad I_{R2} = \text{current through } R_2, \text{etc.}$$
$$V_b = V_1$$
$$V_g = V_3$$

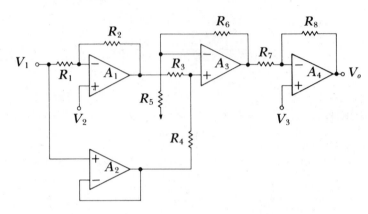

$V_1 = +5$ V dc	$R_1 = 10$ kΩ	$R_4 = 30$ kΩ	$R_7 = 5$ kΩ
$V_2 = +4$ V dc	$R_2 = 20$ kΩ	$R_5 = 10$ kΩ	$R_8 = 10$ kΩ
$V_3 = +5$ V dc	$R_3 = 10$ kΩ	$R_6 = 22$ kΩ	

Fig. 1-29 Operational amplifier configuration of Prob. 15.

a. Do A_1 first:

$$\frac{V_1 - V_a}{R_1} = \frac{V_1 - V_2}{R_1} = I_{R1}$$
$$V_c = V_a - I_{R1}R_2 = V_2 - I_{R1}R_2$$

Numerically:

$$I_{R1} = \frac{5\text{ V} - 4\text{ V}}{10\text{ k}\Omega} = 10^{-4}\text{ A}$$

$$V_c = 4\text{ V} - (10^{-4}\text{ A})(20\text{ k}\Omega) = 2\text{ V dc}$$

b. Do A_2. A_2 is simply a buffer amplifier, and $V_b = V_1$.

c. Do A_3. (*Note*: V_f is the independent voltage.)

$$V_f = \frac{(V_c - V_b)R_4}{R_3 + R_4} + V_b$$

$$= \frac{(2-5)30 \text{ k}\Omega}{40 \text{ k}\Omega} + 5$$

$$= -2.25 + 5 = +2.75 \text{ V dc}$$

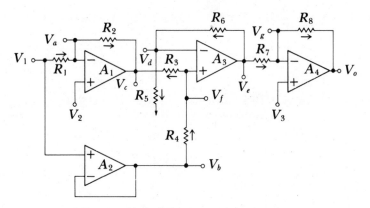

Fig. 1-30   Node-defined amplifier of Prob. 15.

Now as before:

$$V_d = V_f = 2.75$$

$$I_{R6} = \frac{V_d}{R_5} = I_{R5} \qquad \text{[since the current is zero into } (-) \text{ input of } A_3]$$

$$\therefore \quad V_e = V_d + I_{R5}R_6$$

$$= V_d + \frac{V_d R_6}{R_5}$$

$$= 2.75 + \frac{(2.75)(22 \text{ k}\Omega)}{10 \text{ k}\Omega}$$

$$= +8.8 \text{ V dc}$$

d.   Do A_4:

$$V_g = V_3 = +5 \text{ V dc}$$

$$I_{R7} = \frac{V_e - V_g}{R_7} = \frac{8.8 - 5}{5 \times 10^3} = 0.76 \text{ mA}$$

and $I_{R8} = I_{R7}$ [since the current is zero into ($-$) input of A_4]

$$V_o = V_g - I_{R8}R_8$$
$$= 5 - 0.76 \text{ mA}(10 \text{ k}\Omega) = -2.6 \text{ V dc}$$

e. What happens if we change R_6 to 40 kΩ? From part c,

$$V_e = 2.75 + \frac{40}{10}(2.75) = +13.75 \text{ V dc}$$

But this cannot happen. The swing on the output of the operational amplifier limits the positive voltage extreme to +10 V dc.

CHECK THE SOLUTION: Equation (C3-1) of Appendix C3 can be used to check each amplifier section if the appropriate values are substituted.

$$E_o = \frac{(V_1 - V_2)(R_3 + R_4)R_2}{(R_1 + R_2)R_3} + \frac{(R_3 + R_4)V_2}{R_3} - \frac{V_3 R_4}{R_3} \quad \text{(C3-1)}$$

For A_1,

Figure C3-1 designation	Figure 1-30 designation	Value
V_1	V_2	4 V dc
V_2	0	0 V dc
V_3	V_1	5 V dc
R_1		0
R_2		∞
R_3	R_1	10 kΩ
R_4	R_2	20 kΩ
E_o	V_c	

Divide the numerator and the denominator of the first term of Eq. (C3-1) by R_2 and solve for V_c:

$$V_c = \frac{(4-0)(30 \times 10^3)}{10 \times 10^3} + 0 - \frac{5(20 \times 10^3)}{10 \times 10^3} = 2 \text{ V dc}$$

and this checks.

For A_3,

Figure C3-1 designation	Figure 1-30 designation	Value
V_1	V_c	2 V dc
V_2	V_b	5 V dc
V_3	0	0
R_1	R_3	10 kΩ
R_2	R_4	30 kΩ
R_3	R_5	10 kΩ
R_4	R_6	22 kΩ
E_o	V_e	

$$V_e = \frac{(2-5)(32 \times 10^3)(30 \times 10^3)}{(40 \times 10^3)(10 \times 10^3)} + \frac{32 \times 10^3(5)}{10 \times 10^3} - 0 = 8.8 \text{ V dc}$$

and this checks.
   For A_4,

Figure C3-1 designation	Figure 1-30 designation	Value
V_1	V_3	5 V dc
V_2	V_3	5 V dc
V_3	V_e	8.8 V dc
R_1	0	0
R_2	0	0
R_3	R_7	5 kΩ
R_4	R_8	10 kΩ
E_o	V_o	

$$V_o = \frac{(5-5)(15 \times 10^3)(0)}{(0)(5 \times 10^3)} + \frac{(15)(5)}{5 \times 10^3} - \frac{(8.8)(10 \times 10^3)}{5 \times 10^3}$$
$$= -2.6 \text{ V dc}$$

and this checks our solution.

PROBLEM 16

GIVEN: Two points on a semiconductor diode specification. It is desired to find a third point.

FIND: V_D at 5 mA if $V_D = 0.64$ V at 1 mA and 0.7 V at 10 mA. (Refer to Fig. 1-31.)

SOLUTION: The diode equation is given as

$$I_D = I_s \left(\exp \frac{qV}{nKT} - 1\right) = I_s(e^{KV} - 1)$$

where I_s and K are constants determined by the device. I_s is very

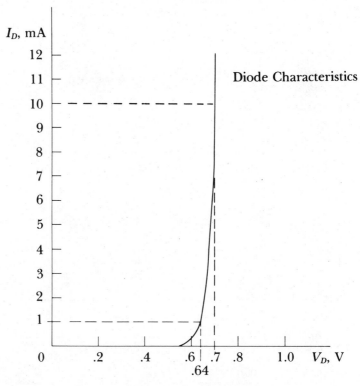

Fig. 1-31   Diode characteristics of Prob. 16.

small. At current levels much larger than I_s (say greater than 1 μA), the equation becomes

$$I_D = I_s e^{KV}$$

and

$$I_s = I_D e^{-KV}$$

Since we have two equations and two unknowns, we can solve for K and I_s:

At $I_D = 1$ mA: $I_s = 10^{-3}e^{-0.64K}$

At $I_D = 10$ mA: $I_s = 10^{-2}e^{-0.7K}$

Solving, we find

$$I_s = 10^{-3}e^{-0.64K} = 10^{-2}e^{-0.7K}$$

and

$$e^{0.06K} = 10$$
$$0.06K = \ln 10$$
$$K = 38.38$$

Then

$$I_s = 10^{-3}e^{-0.64K} = 2.154 \times 10^{-14}$$

Note: Our original assumption that $I_D \gg I_S$ is valid.

Now let $I_D = 5$ mA. Since $I_D = I_s e^{KV}$,

$$\ln\left(\frac{I_D}{I_s}\right) = KV$$

$$V = \frac{\ln\left(\dfrac{I_D}{I_s}\right)}{K} = 0.682 \text{ V}$$

CHECK THE SOLUTION:

$$I_D = I_s e^{KV}$$
$$= 2.154 \times 10^{-14}e^{(38.38)(0.682)}$$
$$= 5.02 \times 10^{-3} \qquad \text{(checks)}$$

PROBLEM 17

GIVEN: The amplifier circuit shown in Fig. 1-32. The load is fixed at $10\,\Omega$ and the output level can vary between 0 and $+10$ V dc, and remain at any one voltage for an indefinite period of time. We desire to design a heat sink for the transistor and need to know what is the maximum power dissipated in Q_1.

FIND: Derive the expression for maximum power in Q_1 and calculate its magnitude.

SOLUTION: Let's find the expression for the power in Q_1 as a function of V_o and take the derivative with respect to V_o to find the maximum power.

$$I_{C1} \approx I_{RL} = \frac{V_o}{R_L}$$

$$V_{CE1} = V_{CC} - V_o$$

$$P_{Q1} = V_{CE1}I_{C1}$$

$$= \frac{(V_{CC} - V_o)V_o}{R_L}$$

$$\frac{dP_{Q1}}{dV_o} = \frac{V_{CC}}{R_L} - \frac{2V_o}{R_L}$$

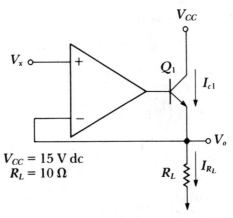

Fig. 1-32   Amplifier configuration of Prob. 17.

Set the derivative = 0 and solve:

$$\frac{V_{CC}}{R_L} - \frac{2V_o}{R_L} = 0$$

$$V_o = \frac{V_{CC}}{2}$$

and

$$P_{Q1} = \frac{(15 - 7.5)(7.5)}{10}$$

$$= 5.625 \text{ W}$$

CHECK THE SOLUTION: Find the power on either side of 7.5 V to see if this solution is correct:

At $V_o = 7.4$ V dc: $P_{Q1} = \dfrac{(15 - 7.4)(7.4)}{10} = 5.624$ W

At $V_o = 7.6$ V dc: $P_{Q1} = \dfrac{(15 - 7.6)(7.6)}{10} = 5.624$ W

Our solution is correct since the power is less on either side of 7.5 V dc.

PROBLEM 18

GIVEN: The circuit shown in Fig. 1-33, which is used in a gain-control application. The input signal level e_i varies from 0 to +2 V dc. We desire to control the gain e_o/e_i from 0.4 to 1.

FIND: Assuming that V_{AGC} controlling the JFET can bias the gate from full on to full off, can we get the control range we need from the JFET with the parameters shown?

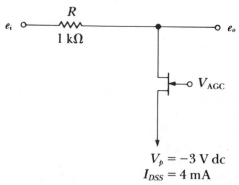

$V_p = -3$ V dc
$I_{DSS} = 4$ mA

Fig. 1-33   AGC circuit of Prob. 18.

SOLUTION: If the JFET is turned fully off, we can assume that we get a gain of 1. We must turn the JFET on, which now becomes the condition to be concerned with. What value of JFET resistance (r_{DS}) do we need?

$$\frac{r_{DS}}{r_{DS} + R} = \frac{e_o}{e_i} = 0.4$$

Solving, we find

$$r_{DS} = 667 \ \Omega$$

Can we get this value? Let's determine the limits of r_{DS} for the values of V_p and I_{DSS} given. The maximum value of r_{DS} can be determined as shown in Fig. 1-34.

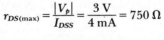

$$r_{DS(max)} = \frac{|V_p|}{I_{DSS}} = \frac{3 \text{ V}}{4 \text{ mA}} = 750 \ \Omega$$

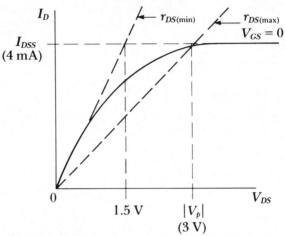

Fig. 1-34   Determining $r_{DS(max)}$ for the JFET of Prob. 18.

The minimum value of r_{DS} is given as r_{DS0} in Appendix F2 as Eq. (F2-3):

$$r_{DS0} = r_{DS(min)} = \frac{0.5|V_p|}{I_{DSS}} = \frac{0.5(3)}{4 \text{ mA}} = 375 \ \Omega \qquad \text{(F2-3)}$$

So, the value of r_{DS} will be between 375 and 750 Ω. Let's calculate the exact value. When $A_v = 0.4$, there is 0.8 V across the JFET (2×0.4) and $V_{DS} = 0.8$ V dc. Also, $V_{GS} = 0$ V dc. Solve for I_D using Eq. (F2-1).

$$I_D = \frac{2 V_{DS} I_{DSS}}{V_p} \left(\frac{V_{GS}}{V_p} - \frac{V_{DS}}{2V_p} - 1 \right) \qquad \text{(F2-1)}$$

$$= \frac{2(0.8)(4 \text{ mA})}{-3} \left(0 - \frac{0.8}{-6} - 1 \right)$$

$$= 1.85 \text{ mA}$$

and r_{DS} at this value of I_D and V_{DS} is

$$r_{DS} \leq \frac{0.8 \text{ V}}{1.85 \text{ mA}} = 432 \ \Omega$$

which meets our requirements because we can get as low as 432 Ω (see Fig. 1-35). We would not turn the JFET on quite so hard, since we only need 667 Ω. If we had been given a range of V_p and I_{DSS}, we would use the maximum value of V_p and minimum value of I_{DSS}.

CHECK THE SOLUTION: This problem does not lend itself well to checking; however, on the examination all numerical answers should be checked. *Note*: As the drain-to-source voltage gets larger, the distortion increases. This was not asked for in the problem.

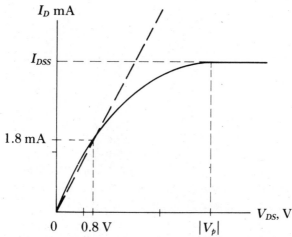

Fig. 1-35   Determining r_{DS} at $I_D = 1.8$ mA for the JFET of Prob. 18.

PROBLEM 19

GIVEN: The circuit of Fig. 1-36 with the circuit values as shown.

FIND:

   a.   The input impedance as seen at the emitter of Q_1.

   b.   The current gain i_c/i_e.

   c.   The voltage gain e_o/e_s.

   d.   The output impedance as seen at the collector.

   Assume that the capacitors are short circuits at the frequency of interest.

SOLUTION: First draw the linear ac equivalent circuit as shown in Fig. 1-37. This is a CB amplifier as described in Appendix A_1. For this circuit, $R_C = R_1 = 4.7$ kΩ.

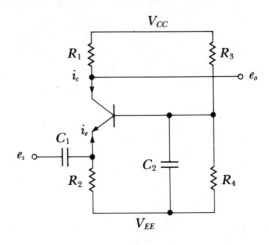

$R_1 = 4.7$ kΩ	$h_{ib} = 10\ \Omega$	$C_1 = 22\ \mu F$
$R_2 = 100\ \Omega$	$h_{ob} = 10^{-6}$ S	$C_2 = 47\ \mu F$
$R_3 = 10$ kΩ	$h_{rb} = 5 \times 10^{-4}$	
$R_4 = 1$ kΩ	$h_{fb} = -.99$	

Fig. 1-36   Circuit configuration of Prob. 19.

a.   First find Δh_b, then Z'_{in}.

$$\Delta h_b = h_{ib}h_{ob} - h_{fb}h_{rb} \qquad\qquad\qquad\text{(A1-10)}$$
$$= (10)(10^{-6}) - (-0.99)(5 \times 10^{-4})$$
$$= 10^{-5} + 4.95 \times 10^{-4} = 5.05 \times 10^{-4}$$

$$Z_{in} = \frac{h_{ib} + R_C\Delta h_b}{1 + R_C h_{ob}} \qquad (R_C = R_1)$$

$$= \frac{10 + 4.7 \times 10^3 \times 5.05 \times 10^{-4}}{1 + 4.7 \times 10^3 \times 10^{-6}} = 12.316\ \Omega \qquad \text{(A1-6)}$$

The Z_{in} just calculated does not include R_2. R_2 is in parallel with Z_{in}, which gives us Z'_{in}, which the problem requests.

$$Z'_{in} = Z_{in}\|R_2 = 12.316\|100$$
$$= 10.97\ \Omega$$

b.   Next determine the current gain A_i.

$$A_i = \frac{i_c}{i_e} = \frac{h_{fb}}{1 + R_C h_{ob}} = \frac{-0.99}{1 + 4.7 \times 10^3 \times 10^{-6}} \qquad \text{(A1-7)}$$

$$= -0.9854$$

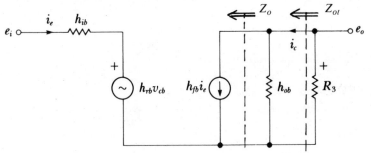

Fig. 1-37   Linear equivalent circuit of Prob. 19.

c.   The voltage gain is

$$A_v = \frac{-h_{fb} R_C}{h_{ib} + R_C \Delta h_b} \qquad \text{(A1-5)}$$

$$= \frac{(0.99)(4.7 \times 10^3)}{10 + (4.7 \times 10^3)(5.05 \times 10^{-4})} = 376$$

d.   The output impedance looking into the collector is

$$Z_{ot} = R_C \| Z_o = R_C \left\| \left(\frac{h_{ib} + R_g}{\Delta h_b + R_g h_{ob}} \right) \right. \qquad \text{(A1-8), (A1-9)}$$

$$= R_C \left\| \frac{h_{ib}}{\Delta h_b} \right. \qquad \text{since } R_g = 0$$

$$= 4.7 \times 10^3 \left\| \left(\frac{10}{5.05 \times 10^{-4}} \right) \right. = 3.8 \text{ k}\Omega$$

CHECK THE SOLUTION: Assume 1 mV for e_{in}.

$$e_o = A_v e_{in} = 376 \text{ mV}$$

$$i_e = \frac{e_{in} - h_{rb} V_{cb}}{h_{ib}} = 81.2 \ \mu\text{A}$$

$$i_c = \frac{e_o}{R_c} = 80 \ \mu\text{A} \qquad A_i = \frac{i_c}{i_e} = 0.9852$$

Another check:

$$i_c = h_{fb}i_e + h_{ob}V_{ce} = -80.014 \qquad \text{(checks with previous)}$$

To check output impedance, load the output with Z_{ot} and the gain should go to one-half its previous value.

$$R_c' = R_c \| Z_{ot} = 4.7 \text{ k}\Omega \| 3.8 \text{ k}\Omega = 2.1 \text{ k}\Omega$$

$$A_v' = \frac{-h_{fb}R_c'}{h_{ib} + R_c' \Delta h_b} = 188$$

which is half the original gain.

PROBLEM 20

GIVEN: The circuit of Fig. 1-38 and the parameters as shown.

FIND:

    a.   The voltage gain e_o/e_{in}.

    b.   The input impedance of the amplifier as seen at the base.

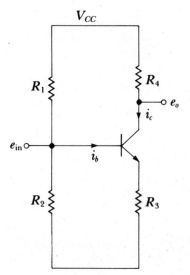

$V_{CC} = +40$ V dc	$h_{fe} = 50$	$R_2 = 10$ kΩ
$h_{ie} = 1$ kΩ	$h_{re} = 10^{-4}$	$R_4 = 7.5$ kΩ
$h_{oe} = 10^{-5}$	$R_1 = 100$ kΩ	$R_3 = 1$ kΩ

Fig. 1-38   Amplifier circuit of Prob. 20.

c.   The current gain i_c/i_b.

d.   The output impedance as seen at the collector.

SOLUTION:

a.   The voltage gain can be found by using the equations of Appendix A5. Draw the linear ac equivalent circuit and then calculate the gain. The linear ac equivalent circuit is shown in Fig. 1-39. R_1 and R_2 do not enter into the problem because the voltage is applied to the base and the internal impedance of e_{in} is assumed zero. The gain can be found from Appendix A5 as follows:

$$A_v = \frac{e_o}{e_{in}} = \frac{(h_{oe}R_E - h_{fe})R_C}{(h_{ie} + R_C\Delta h_e')(1 + h_{oe}R_E) + (1 - h_{re})(1 + h_{fe})R_E}$$

$$(A5\text{-}5)$$

Solve for $\Delta h_e'$. Note: $R_E = R_3$, $R_C = R_4$.

$$\Delta h_e' = \frac{\left\{ [h_{ie}(1 + h_{oe}R_E) + (1 - h_{re})(1 + h_{fe})R_E]h_{oe} - (h_{re} + h_{oe}R_E)(h_{fe} - h_{oe}R_E) \right\}}{(1 + h_{oe}R_E)^2}$$

$$(A5\text{-}10)$$

$$= \frac{\left\{ [10^3(1 + 10^{-5} \times 10^3) + (1 - 10^{-4})(1 + 50)10^3]10^{-5} - (10^{-4} + 10^{-5} \times 10^3)(50 - 10^{-5} \times 10^3) \right\}}{(1 + 10^{-5}10^3)^2}$$

$$= \frac{0.520 - 0.5049}{(1.02)} = 14.8 \times 10^{-3}$$

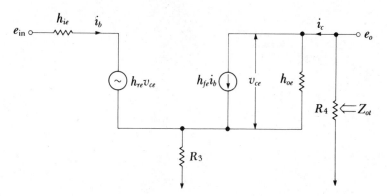

Fig. 1-39   Linear equivalent circuit of Prob. 20.

$$A_v = \frac{(10^{-5} \times 10^3 - 50)7.5 \times 10^3}{\left[\begin{array}{c} (10^3 + 7.5 \times 10^3 \times 14.8 \times 10^{-3})(1 + 10^{-5} \times 10^3) \\ + (1 - 10^{-4})(1 + 50)10^3 \end{array} \right]}$$

$$= \frac{-3.75 \times 10^5}{1.111 \times 10^3 + 50.995 \times 10^3}$$

$$= -7.197$$

b. The input impedance can be found from Eq. (A5-6):

$$Z_{in} = \frac{(1 + h_{oe}R_E)(h_{ie} + R_C\Delta h_e') + (1 - h_{re})(1 + h_{fe})R_E}{1 + (R_C + R_E)h_{oe}} \qquad \text{(A5-6)}$$

$$= \frac{\left[\begin{array}{c} (1 + 10^{-5} \times 10^3)(10^3 + 7.5 \times 10^3 \times 14.8 \times 10^{-3}) \\ + (1 - 10^{-4})(1 + 50)10^3 \end{array} \right]}{1 + (7.5 \times 10^3 + 10^3)10^{-5}}$$

$$= \frac{1.122 \times 10^3 + 50.995 \times 10^3}{1.085} = 48.03 \text{ k}\Omega$$

c. The current gain can be found from Eq. (A5-7):

$$A_I = \frac{h_{fe} - h_{oe}R_E}{1 + (R_C + R_E)h_{oe}} \qquad \text{(A5-7)}$$

$$= \frac{50 - 10^{-5} \times 10^3}{1 + (7.5 \times 10^3 + 10^3)10^{-5}} = 46.07$$

d. The output impedance can be found from Eqs. (A5-8) and (A5-9). We want Z_{ot}.

$$Z_o = \frac{(h_{ie} + R_g)(1 + h_{oe}R_E) + (1 - h_{re})(1 + h_{fe})R_E}{\Delta h_e'(1 + h_{oe}R_E) + R_gh_{oe}} \qquad \text{(A5-8)}$$

$$= \frac{(10^3 + 0)(1 + 10^{-5} \times 10^3) + (1 - 10^{-4})(1 + 50)10^3}{14.8 \times 10^{-3}(1 + 10^{-5} \times 10^3) + 0(10^{-5})}$$

$$= \frac{52 \times 10^3}{14.95 \times 10^{-3}} = 3.48 \times 10^6 \ \Omega$$

$$Z_{ot} = Z_o \| R_4 = 3.48 \times 10^6 \| 7.5 \times 10^3 \qquad \text{(A5-9)}$$

$$= 7.48 \text{ k}\Omega$$

CHECK THE SOLUTION: Assume 1 mV input (e_{in}) to the circuit.

$$e_0 = Ae_{in} = -7.197 \text{ mV}$$

$$i_b = \frac{e_{in}}{Z_{in}} = 20.82 \times 10^{-9} \text{ A}$$

$i_c = 0.96 \ \mu A$; $i_e = 0.981 \ \mu A$; $V_e = 0.981$ mV; $V_{ce} = e_o - v_e = -8.18$ mV; $h_{re}V_{ce} = -8.18 \times 10^{-7}$ V; $V_x = V_e + h_{re}V_{ce} = -0.98$ mV; $i_b = 20 \times 10^{-9}$ A (vs. 20.82×10^{-9}). $i_c = h_{fe}i_b + V_{ce}h_{oe} = 0.9592 \ \mu A$ (vs. $0.959 \ \mu A$). The output impedance can be checked by loading the output with a value equal to Z_{ot}. This will cause the gain to go to one-half its previous value. $R'_c = Z_{ot}\|R_4 = 3.745$ kΩ. $A'_v = 3.596$, which is 0.499 times the original gain.

PROBLEM 21

GIVEN: The circuit of Fig. 1-40.

$I_{DSS} = 4$ mA	$R_1 = 100$ kΩ
$V_p = -6$ V	$R_2 = 15$ kΩ
$V_{DD} = +30$ V dc	$R_S = 1$ kΩ
$V_{SS} = -20$ V dc	$R_D = 4.7$ kΩ

FIND: V_S, V_D, V_G, I_D; neglect I_{GSS}.

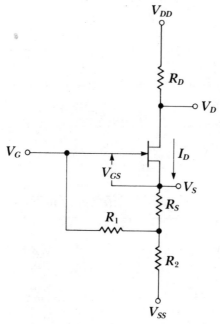

Fig. 1-40 JFET amplifier of Prob. 21.

SOLUTION: Let's get the circuit into the general form of Eq. (F1-10) and solve for I_D using Eq. (F1-11). Since $I_{GSS} = 0$, $I_{R1} = 0$; and V_G is the same voltage that exists at the junction of R_S, R_1, and R_2. The defining equations are:

$$V_G = V_{SS} + I_D R_2$$

$$V_S = I_D(R_S + R_2) + V_{SS}$$

Solving, we find

$$V_{GS} = V_G - V_S = V_{SS} + I_D R_2 - I_D(R_S + R_2) - V_{SS} = -I_D R_S$$

$$I_D = I_{DSS}\left(1 - \frac{V_{GS}}{V_p}\right)^2 \tag{F1-1}$$

Substituting for V_{GS} we find

$$I_D = I_{DSS}\left(1 + \frac{I_D R_S}{V_p}\right)^2$$

which is in the form of Eq. (F1-10):

$$I_D = I_{DSS}(\alpha + \beta I_D)^2 \tag{F1-10}$$

Therefore

$$\alpha = 1 \qquad \beta = \frac{R_S}{V_p} = \frac{1\,\text{k}\Omega}{-6\,\text{V}} = -167$$

Substituting into Eq. (F1-11), we can solve for I_D:

$$I_D = -\frac{\left(\frac{2\alpha}{\beta} - \frac{1}{\beta^2 I_{DSS}}\right) \pm \left[\left(\frac{2\alpha}{\beta} - \frac{1}{\beta^2 I_{DSS}}\right)^2 - \frac{4\alpha^2}{\beta^2}\right]^{1/2}}{2} \tag{F1-11}$$

$$= \frac{\left(-\left[\frac{2}{-166.7} - \frac{1}{(-166.7)^2(4\times10^{-3})}\right] \pm \left\{\left[\frac{2}{(-166.7)} - \frac{1}{(-166.7)^2(4\times10^{-3})}\right]^2 - \frac{4}{(-166.7)^2}\right\}^{1/2}\right)}{2}$$

$$= -\frac{(-12\times10^{-3} - 9\times10^{-3}) \pm (4.41\times10^{-4} - 1.44\times10^{-4})^{1/2}}{2}$$

$$= \frac{21\times10^{-3} \pm 17.233\times10^{-3}}{2} = 19.12\,\text{mA},\ 1.883\,\text{mA}$$

Is $I_D = 19.12$ mA? (This value is not acceptable since it is larger than I_{DSS}.) Check using Eq. (F1-1) for $I_D = 1.883$ mA:

$$I_D = I_{DSS}\left(1 + \frac{I_D R_S}{V_p}\right)^2 = 4 \times 10^{-3}\left[1 + \frac{(1.883 \times 10^{-3})(1 \times 10^3)}{-6}\right]^2$$
$$= 1.883 \text{ mA}$$

This checks!

Now solve for the bias voltages;

$$V_D = V_{DD} - I_D R_D$$
$$= 30 \text{ V} - (1.883 \times 10^{-3})(4.7 \times 10^3)$$
$$= 30 - 8.85 = 21.15 \text{ V dc}$$

$$V_G = V_{SS} + I_D R_2$$
$$= -20 + (1.883 \times 10^{-3})(15 \times 10^3)$$
$$= 8.245 \text{ V dc}$$

$$V_S = V_G + I_D R_S$$
$$= 8.245 + (1.883 \times 10^{-3})(1 \times 10^3)$$
$$= 10.128 \text{ V dc}$$

CHECK THE SOLUTION:

$$V_{GS} = V_G - V_S = 8.245 - 10.128 = -1.883 \text{ V dc}$$
$$V_{GS} = -I_D R_S = -(1.883 \text{ mA})(1 \text{ k}\Omega) = -1.883 \text{ V dc}$$

Since they are equal, the solution checks.

As described in Appendix F, we know that V_{DS} must be larger than $|V_p|$. $|V_p| = 6$ V. What is V_{DS}?

$$V_{DS} = V_D - V_S = 21.15 - 10.128 = 11.022 \text{ V}$$

and our check is complete.

PROBLEM 22

GIVEN: The input circuit of Fig. 1-41, which represents that of a hybrid-π CE amplifier.

FIND: The expression that determines the value of C_x, and explain your results. Assume $f \ll 1/(2\pi R_L C_u)$.

SOLUTION: Let's use the hybrid-π model given in Appendix A8. We neglect r_μ since it is assumed to be much larger than X_{C_μ} at the higher

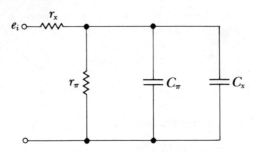

Fig. 1-41 Hybrid-π input circuit of Prob. 22.

frequencies. Let's derive the expression for the impedance to the right of the dashed line in Fig. 1-42 in the hope that we can extract a capacitance term from the expression. Let's solve for i_μ/V:

$$V = i_\mu \left(\frac{1}{sC_\mu}\right) + (i_\mu - g_m V)R_L$$

$$= i_\mu \left(\frac{1}{sC_\mu} + R_L\right) - g_m VR_L$$

$$V(1 + g_m R_L)sC_\mu = i_\mu(1 + R_L C_\mu s) \qquad \text{for } f \ll 1/(2\pi R_L C_\mu)$$

$$\frac{i_\mu}{V} = (1 + g_m R_L)sC_\mu$$

and

$$C_x = (1 + g_m R_L)C_\mu$$

What is the significance of this? Most engineers familiar with transistor amplifiers recognize C_x as the Miller-effect capacitance whereby the collector-base capacitance (C_{ob} or an external collector-base capacitor)

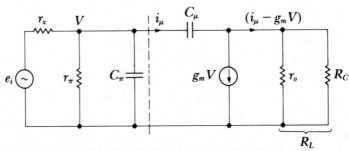

Fig. 1-42. Hybrid-π model from Appendix A8.

is multiplied by 1 plus the gain $(g_m R_L)$. The reason this is true is that the input and output are out of phase, and as the input current is trying to charge the collector-base capacitance, the output is working against it by increasing the voltage across the capacitance in a manner to increase the charge required.

PROBLEM 23

GIVEN: The three-stage amplifier of Fig. 1-43 with its parameters listed. The equivalent shunt capacitance from gate to ground is 100 pF for each stage. Assume C_2 does not limit the low-frequency response.

FIND:

   a.   The total midband gain in decibels (dB).

   b.   The low-frequency breakpoint f'_L for each stage.

   c.   The high-frequency breakpoint f'_H for each stage.

   d.   The overall lower 3-dB frequency f_L.

   e.   The overall upper 3-dB frequency f_H.

SOLUTION:

   a.   The gain expression for midband gain can be found in Appendix B1. At midband, we will assume X_{C1} and X_{C2} are short circuits and check later. Draw the equivalent circuit for one stage as shown in Fig. 1-44.

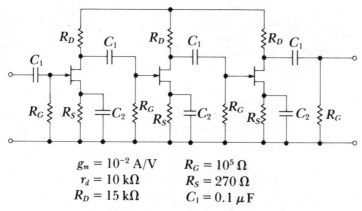

$$g_m = 10^{-2}\,\text{A/V} \qquad R_G = 10^5\,\Omega$$
$$r_d = 10\,\text{k}\Omega \qquad R_S = 270\,\Omega$$
$$R_D = 15\,\text{k}\Omega \qquad C_1 = 0.1\,\mu\text{F}$$

Fig. 1-43   Three-stage amplifier of Prob. 23.

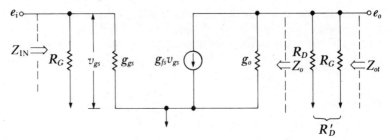

Fig. 1-44 One-stage equivalent circuit of amplifier of Prob. 23.

$$A_v = \frac{e_o}{e_{in}} = \frac{-g_{fs}R_D}{1 + g_o(R_D + R_S) + g_{fs}R_S} \qquad \text{(B1-1)}$$

For this problem, $g_m = g_{fs} = 10^{-2}$, $g_o = 1/r_d = 10^{-4}$, and R_D of Eq. (B1-1) will be relabeled R_D', which equals $R_D \| R_G$.

$$R_D' = R_D \| R_G = 15 \text{ k}\Omega \| 100 \text{ k}\Omega = 13.043 \text{ k}\Omega$$

$$A_v = \frac{-(10^{-2})(13.043 \times 10^3)}{1 + (10^{-4})(13.043 \times 10^3 + 0) + (10^{-2})(0)} = -56.6$$

This gain raised to a power of 3 gives the total gain A_{vt}:

$$A_{vt} = (A_v)^3 = (-56.6)^3 = -1.813 \times 10^5$$

and in decibels,

$$A_{vt} = (20) \log |A_v| = 20 \log (-1.813 \times 10^5) = 105.2 \text{ dB}$$

b, c.   To determine either the low-frequency breakpoint or high-frequency breakpoint, we must derive the expression for the gain as a function of frequency. We have already determined the magnitude of the midband gain. We can use the circuit of Fig. 1-45 to help us determine the low-frequency breakpoint. Neglect R_S since it is shunted by C_2.

$1/g_{gs}$ is assumed to be much larger than R_G, which is a reasonable assumption. Let's solve for $e_o(s)/I(s)$, which will provide us with information to be used in the determination of our 3-dB frequencies only. Let

$$Z_G = R_G \left\| \frac{1}{C_{GS}} = \frac{R_G}{R_G C_{GS}s + 1} = \frac{N}{D}$$

$$Z_2 = \frac{1}{C_1 s} + Z_G = \frac{R_G C_1 s + 1}{(R_G C_G s + 1)C_1 s} = \frac{N_2}{D_2}$$

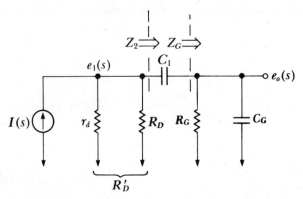

Fig. 1-45   The ac equivalent circuit of one-stage of Prob. 23.

$$Z_1 = R_D' \| Z_2 = \frac{R_D' N_2}{R_D' D_2 + N_2}$$

$$e_1(s) = I(s)Z_1$$

$$\frac{e_1(s)}{I(s)} = Z_1$$

$$\frac{e_o(s)}{e_1(s)} = \frac{Z_G}{Z_2} = \frac{N/D}{N_2/D_2}$$

$$\frac{e_o(s)}{I(s)} = \frac{e_o(s)}{e_1(s)} \frac{e_1(s)}{I(s)} = \frac{N/D}{N_2/D_2} \frac{R_D' N_2}{R_D' D_2 + N_2}$$

$$= \frac{N D_2 R_D'}{D(R_D' D_2 + N_2)} = \frac{R_G C_1 s (R_G C_G s + 1) R_D'}{(R_G C_G s + 1)[R_D' C_1 s (R_G C_G s + 1) + R_G C_1 s + 1]}$$

$$= \frac{R_G R_D' C_1 s}{R_D' R_G C_1 C_G s^2 + [R_D' C_1 + R_G (C_1 + C_G)]s + 1}$$

which is of the form

$$\frac{e_o(s)}{I(s)} = \frac{T_1 s}{(T_3 s + 1)(T_4 s + 1)}$$

Therefore

$$T_1 = R_G R_D' C_1 = 13$$

$$T_3 T_4 = R_D' R_G C_1 C_G = 1.3 \times 10^{-9}$$

$$T_3 + T_4 = R_D' C_1 + R_G (C_1 + C_G) = R_D' C_1 + R_G C_1 = 1.14 \times 10^{-3}$$

Solving these equations simultaneously we find

$$T_3 = 1.14 \times 10^{-3}$$

$$T_4 = 1.15 \times 10^{-6}$$

and our overall transfer function becomes

$$\frac{e_o(s)}{I(s)} = \frac{13s}{(1.14 \times 10^{-3}s + 1)(1.15 \times 10^{-6}s + 1)} = \frac{9.92 \times 10^{10}}{(s + 877)(s + 8.7 \times 10^5)}$$

$$f_L' = f_3 = 1/(2\pi T_3) \cong 139.6 \text{ Hz}$$
$$f_H' = f_4 = 1/(2\pi T_4) = 138.5 \text{ kHz}$$

This gives us the Bode plot shown in Fig. 1-46.

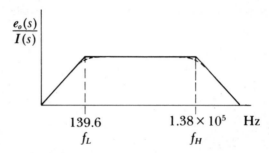

Fig. 1-46   Bode plot of one-stage of Prob. 23.

c.   The overall low-frequency breakpoint can be determined by the equation given in Appendix D4.

$$f_L = \frac{f_L'}{(2^{1/N} - 1)^{1/2}} = \frac{139.6}{0.51} \text{ Hz} = 273.8 \text{ Hz} \qquad \text{(D4-2)}$$

d.   Also, in Appendix D9 we find the equation for the equivalent high-frequency bandwidth.

$$f_H = f_H'(2^{1/N} - 1)^{1/2} = 70.64 \text{ kHz} \qquad \text{(D4-1)}$$

CHECK THE SOLUTION: From (D4-3), $f_{mid} = (f_L' f_H')^{1/2} = 4.4$ kHz and $X_{c1} = 362 \ \Omega$, which is small compared with 100 kΩ.

$$g_{fs}V_{gs} = \frac{-e_o}{R_D \| R_G \| \dfrac{1}{g_o}} = 10^{-5}$$

$$A_v = -10 \exp \left[\frac{(105.2 \text{ dB})}{20} \right] = -10^{5.26} = -1.8192 \times 10^5$$

(vs. -1.813×10^5)

PROBLEM 24

GIVEN: The circuit shown in Fig. 1-47.

FIND:

a.   Value of R_f for a transfer impedance of e_o/i_{in} of -10^6.

b.   If $R_f = 10 \text{ k}\Omega$, find $A_v = e_o/e_{in}$.

SOLUTION:

a.   This problem could become quite complex if we don't make two simplifying assumptions: $R_f \gg R_3$ and $R_f \gg R_i$, which are reasonable for most practical problems. If we do this, we open-circuit R_f and have the equivalent open-loop circuit of Fig. 1-48. We then can calculate the equivalent open-loop gain and use conventional feedback theory to determine the closed-loop response. The other way of solving the problem is to assume that the open-loop gain is large enough that it can be considered to be ∞.

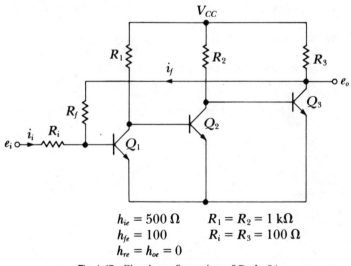

$h_{ie} = 500 \ \Omega$ $R_1 = R_2 = 1 \text{ k}\Omega$
$h_{fe} = 100$ $R_i = R_3 = 100 \ \Omega$
$h_{re} = h_{oe} = 0$

Fig. 1-47   Circuit configuration of Prob. 24.

Assume the former to be the case, and solve for the open-loop gain, input impedance, and output impedance before solving for the transfer impedance and gain. The three stages of gain are all CE amplifiers and we can use the gain and impedance equations of Appendix A2. For the overall gain,

$$\frac{e_o'}{e_{in}} = \left(\frac{e_o'}{e_c}\right)\left(\frac{e_c}{e_b}\right)\left(\frac{e_b}{e_a}\right)$$

For all amplifiers, $\Delta h_e = h_{ie}h_{oe} - h_{fe}h_{re} = 0$; $R_{C3} = R_3$, and

$$\frac{e_o'}{e_c} = \frac{-h_{fe,3}R_{C3}}{h_{ie,3} + R_C\Delta h} = \frac{-h_{fe,3}R_{C3}}{h_{ie,3}} = \frac{-(100)(100)}{500} = -20 \quad (A2\text{-}5)$$

The input impedance of Q_3 is $h_{ie,3}$. The equivalent collector load of Q_2 is $R_2\|h_{ie,3} = R_{C2}$.

$$R_{C2} = 1\ \text{k}\Omega\|500 = 333\ \Omega$$

$$A_{v2} = \frac{e_c}{e_b} = \frac{-h_{fe,3}R_{C3}}{h_{ie,2}} = \frac{-(100)(333)}{500} = -66.7$$

The input impedance of Q_2 is the same as that of Q_3. Therefore R_{C1} is the same as R_{C2} and $A_{v1} = A_{v2} = e_b/e_a = -66.7$, and our total open-loop gain is

$$A_o = \frac{e_o'}{e_{in}} = (-20)(-66.7)(-66.7) = -8.9 \times 10^4$$

The minus sign is good because it tells us we have negative feedback. Now, we can use the equations of Appendix C2 for our equivalent closed-loop gain, A_v. Refer to Fig. 1-49 for our new

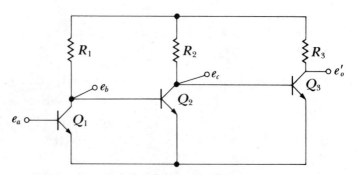

Fig. 1-48 Open-loop representation of circuit of Prob. 24.

equivalent circuit. The input impedance of the amplifier of Fig. 1-49 is nothing more than the input impedance of Q_1, which is $h_{ie,1} = 500\ \Omega$. The output impedance of the amplifier is the output impedance Z_{ot} of the CE stage of Q_3. Equation (A2-8) tells us that Z_o is $\infty\ \Omega$ since $h_{re} = h_{oe} = 0$, and $Z_{ot} = R_3 = 100\ \Omega$. Therefore, we can use the following relationships for the circuit of C2-1.

Figure C2-1	Figure 1-49	Value
R_{in}	$h_{ie,1}$	$500\ \Omega$
A_o	A_o	-8.9×10^4
R_o	R_o	100
Z_i	R_i	100
Z_f	R_f	10^4
Z_s	0	0

$$A_v = \frac{e_o}{e_{in}} = \frac{R_o(R_{in} + Z_s) + A_oR_{in}Z_f}{(Z_f + R_o)(R_{in} + Z_s) + Z_i(Z_f + R_o + R_{in} + Z_s) - A_oR_{in}Z_i}$$
$$\text{(C2-2)}$$

Put into terms of our problem,

$$A_v = \frac{R_o(h_{ie,1} + 0) + A_oh_{ie,1}R_f}{(R_f + R_o)(h_{ie,1} + 0) + R_i(R_f + R_o + h_{ie,1} + 0) - A_oh_{ie,1}R_i}$$

$$= \frac{(100)(500) + (-8.9 \times 10^4)(500)R_f}{\left[\begin{array}{c}(R_f + 100)(500 + 0) + 100(R_f + 100 + 500 + 0) \\ -(-8.9 \times 10^4)(500)(100)\end{array}\right]}$$

$$\approx \frac{5 \times 10^4 - 4.45 \times 10^7 R_f}{600R_f + 4.45 \times 10^9}$$

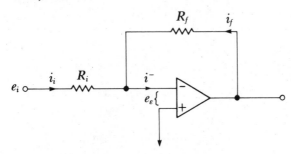

Fig. 1-49 Check circuit for Prob. 24.

Now what value do we want A_v to be for a transfer impedance of 10^6?

$$A_v = \frac{e_o}{e_{in}} = \frac{e_o}{i_{in}R_i}$$

$$\frac{e_o}{i_{in}} = A_v R_i = -10^6$$

$$A_v = \frac{-10^6}{R_i} = -10^4$$

Therefore we can solve for R_f:

$$A_v = \frac{5 \times 10^4 - (4.45 \times 10^7)R_f}{600R_f + 4.45 \times 10^9} = -10^4$$

$$R_f = 1.156 \times 10^6 \ \Omega$$

b.   For $R_f = 10 \ k\Omega$,

$$A_v = \frac{e_o}{e_{in}} = \frac{(100)(500) + (-8.9 \times 10^4)(500)(10 \times 10^3)}{500(10 \times 10^3) + 100(10.1 \times 10^3) - (-8.9 \times 10^4)(500)(100)}$$

$$\approx \frac{-4.45 \times 10^{11}}{4.456 \times 10^9} = -99.865$$

CHECK THE SOLUTION:

a.   Refer to the circuit of Fig. 1-49. First we can approximate the gain as:

$$A_v \approx \frac{-R_f}{R_i} = \frac{-1.15 \times 10^6}{100} = -1.156 \times 10^4$$

which is close to the exact value of $A_v = -10^4$. When the closed-loop gain starts approaching the open-loop gain, the error increases. Now, assume we have 1 mV at the input.

$$e_o = A_v e_{in} = -10^4 \times 10^{-3} = -10 \ V$$

The error input voltage from the negative terminal to ground is this voltage divided by the open loop gain:

$$e_\epsilon = \frac{-10 \ V}{-8.9 \times 10^4} = 0.1124 \ mV$$

Let's use these voltages to check the node currents:

$$i_{in} = \frac{e_{in} - e_\epsilon}{R_i} = \frac{(1 - 0.1124)\,\text{mV}}{100} = 8.876\,\mu\text{A}$$

$$i^- = \frac{e_\epsilon}{h_{ie,1}} = \frac{0.1124\,\text{mV}}{500} = 0.225\,\mu\text{A}$$

$$i_f = -(i_{in} - i^-) = -8.651\,\mu\text{A}$$

$$e_o = e_\epsilon + i_f R_f$$
$$= 0.1124\,\text{mV} - (8.651\,\mu\text{A})(1.156 \times 10^6)$$
$$= -10\,\text{V}$$

$$\frac{e_o}{i_{in}} = A_v R_i = -10^4 \times 10^2 = -10^6$$

which checks this answer; with 1 mV in and a gain of -10^4 we should get -10 V output.

PROBLEM 25

GIVEN: The transistor amplifier of Fig. 1-50.

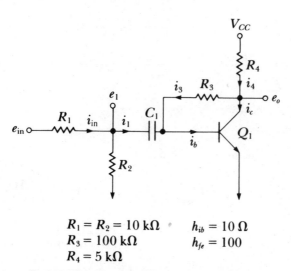

$$R_1 = R_2 = 10\,\text{k}\Omega \qquad h_{ib} = 10\,\Omega$$
$$R_3 = 100\,\text{k}\Omega \qquad h_{fe} = 100$$
$$R_4 = 5\,\text{k}\Omega$$

Fig. 1-50 Transistor amplifier circuit of Prob. 25.

FIND:

   a. i_4/i_{in}

   b. e_o/e_{in}

   c. $Z_{in} = e_1/i_1$

   d. e_o/e_1

SOLUTION: This is a rather involved problem. If we find the voltage gain first, it will simplify our calculations greatly. Let's neglect the capacitor; assume it is a short circuit. We have the equations for this problem in Appendix A6. The linear ac equivalent circuit is shown in Fig. 1-51. We will assume $h_{re} = h_{oe} = 0$ because they are not given. From Appendices A6 and A7 we have (*note*: $R_F = R_3$):

$$h_{ie} = (h_{fe} + 1)h_{ib} = 1.01 \text{ k}\Omega \tag{A2-12}$$

$$h_{ie}^* = \frac{h_{ie}R_F}{h_{ie} + R_F} = (1.01 \times 10^3)\|(100 \times 10^3) \approx 10^3 \ \Omega \tag{A6-10}$$

$$h_{oe}^* = h_{oe} + \frac{(1 - h_{re})(1 + h_{fe})}{h_{ie} + R_F}$$

$$= 0 + \frac{1(101)}{1.01 \times 10^3 + 100 \times 10^3} = 10^{-3} \text{ S} \tag{A6-11}$$

$$h_{fe}^* = \frac{h_{fe}R_F - h_{ie}}{h_{ie} + R_F} = \frac{(10^2)(10^5) - 10^2}{1.01 \times 10^3 + 100 \times 10^3} = 99 \tag{A6-12}$$

$$h_{re}^* = h_{re} + \frac{(1 - h_{re})h_{ie}}{h_{ie} + R_F} = 0 + \frac{h_{ie}}{h_{ie} + R_F} = \frac{1.01 \times 10^3}{101 \times 10^3} = 10^{-2} \tag{A6-13}$$

$$\Delta h^* = h_{ie}^* h_{oe}^* - h_{fe}^* h_{re}^* \tag{A6-14}$$

$$= (10^3)(10^{-3}) - (99)(10^{-2})$$

$$= 0.01$$

The gain e_o/e_1 is the answer to part d:

   d. $$A_v = \frac{e_o}{e_1} = \frac{-h_{fe}^* R_L}{\Delta h^* R_L + h_{ie}^*} = \frac{-(99)(5 \times 10^3)}{(0.01)(5 \times 10^3) + 10^3} = -471.4$$
$$\tag{A6-5}$$

Note that $R_L = R_4$ for this problem. The input impedance as seen at the base of Q_1 is answer c and can be solved for as follows:

   c. $$Z_{in1} = \frac{e_1}{i_1} = \frac{\Delta h^* R_L + h_{ie}^*}{1 + h_{oe}^* R_L} = \frac{(0.01)(5 \times 10^3) + 10^3}{1 + (10^{-3})(5 \times 10^3)} \tag{A6-6}$$

$$= 175 \ \Omega$$

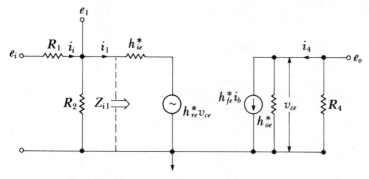

Fig. 1-51   Linear ac equivalent circuit of Prob. 25.

The gain from the input to the base of Q_1 is

$$\frac{e_1}{e_{in}} = \frac{R_2\|Z_{in1}}{R_2\|Z_{in1} + R_1} = \frac{(10 \times 10^3)\|175}{[(10 \times 10^3)\|175] + 10 \times 10^3} = 1.691 \times 10^{-2}$$

The total gain is the answer to part b:

b. $\dfrac{e_o}{e_{in}} = \dfrac{e_0}{e_1} \cdot \dfrac{e_1}{e_{in}} = (-471.4)(1.691 \times 10^{-2}) = -7.971$

This leaves only the current-gain ec.. 'on of part a to be solved. From Appendix A6,

$$A_i = \frac{i_c^*}{i_b^*} = \frac{i_4}{i_1} = \frac{h_{fe}^*}{1 + h_{oe}^* R_L} = \frac{99}{1 + (10^{-3})(5 \times 10^3)} = 16.5 \qquad \text{(A6-7)}$$

Next we can find i_1/i_{in}. The current i_1 is related to i_{in} by the equation

$$i_1 = \frac{R_2 i_{in}}{R_2 + Z_{in1}} = \frac{(10 \times 10^3) i_{in}}{10 \times 10^3 + 175}$$

$$\frac{i_1}{i_{in}} = 0.983$$

Therefore, we can now find the answer to part a:

a. $\dfrac{i_4}{i_{in}} = \dfrac{i_4}{i_1} \cdot \dfrac{i_1}{i_{in}} = (16.5)(0.983) = 16.22$

CHECK THE SOLUTION: Put 1 V in at e_{in};

$$e_o = A_v e_{in} = -7.971 \text{ V} \qquad e_1 = 16.91 \text{ mV}$$

$$i_{in} = \frac{e_{in} - e_1}{R_1} = 98.31 \ \mu\text{A}$$

$$i_1 = \frac{e_1}{Z_{in1}} = 96.63 \ \mu A$$

$$i_2 = \frac{e_1}{R_2} = 1.691 \ \mu A$$

$$i_3 = \frac{e_o - e_1}{R_3} = -79.5 \ \mu A$$

$$i_4 = \frac{0 - e_o}{R_4} = 1.594 \ \text{mA}$$

$$i_{in} = i_1 + i_2 = 98.32 \ \mu A \ (\text{vs. } 98.31)$$

$$\frac{i_4}{i_1} = 16.5 \ (\text{vs. } 16.22)$$

$$i_b = i_1 + i_3 = 17.13 \ \mu A$$

and

$$i_c = i_4 - i_3 = 1673.5 \ \mu A$$

$$h_{fe} = i_c/i_b = 100.03 \ (\text{vs. } 100)$$

PROBLEM 26

GIVEN: A voltage regulator is to be made using a nonlinear element as its reference and a resistor in series (R_s) as shown in Fig. 1-52. The current/voltage relationship for the nonlinear element R_x is $I_x = 0.01(V_x)^{1.2}$.

FIND: Suppose a resistor R_1 is placed in parallel with R_s, and E_o is desired to be 5 V dc with maximum load. What are the values for R_s and R_1 if we want a no-load to full-load regulation of $\pm 5\%$?

SOLUTION: Let $R_p = R_s \| R_1$. Determine   the   nominal   full-load operating current for reference:

$$V_{x(nom)} = E_{in} - E_o = 25 - 5 = 20 \ \text{V dc}$$
$$I_{x(nom)} = 0.01(20)^{1.2} = 364.1 \ \text{mA}$$

Assume that R_1 can vary from no load (∞) to full load. At no load, the current is minimum and E_o is maximum. At maximum load, E_o is minimum. The voltage $V_x = E_{in} - E_o$ is therefore minimum at no load and maximum at full load.

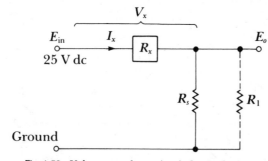

Fig. 1-52 Voltage regulator circuit for Prob. 26.

$I_{x(max)} = 0.01(25 - 4.75)^{1.2} = 0.01(20.25)^{1.2}$

$\qquad = 369.6 \text{ mA} \qquad \text{(full load)}$

$I_{x(min)} = 0.01(25 - 5.25)^{1.2} = 0.01(19.75)^{1.2}$

$\qquad = 358.7 \text{ mA} \qquad \text{(minimum load)}$

$R_{p(min)} = 4.75 \text{ V}/I_{x(max)} = 4.75 \text{ V}/369.6 \text{ mA} = 12.85 \text{ }\Omega$

$R_{p(max)} = R_s = 5.25/I_{x(min)} = 14.64 \text{ }\Omega \qquad \text{(because } R_1 \text{ is open)}$

$1/R_{1(min)} = 1/R_{p(min)} - 1/R_s = 1/12.85 - 1/14.64$

$\quad R_{1(min)} = 105.1 \text{ }\Omega$

ANSWER: $R_1 = 105.1 \text{ }\Omega$; $R_s = 14.64 \text{ }\Omega$.

CHECK THE SOLUTION: The most useful check is to check each step numerically.

PROBLEM 27

GIVEN: The circuit of Fig. 1-53. This is the front end of a discrete operational amplifier. Because of the internal biasing structure of the amplifier, V_E varies from $+2$ V dc to $+6$ V dc.

FIND: The input differential offset voltage change due to the mismatch in emitter currents of Q_1 and Q_2, which are an identically matched pair.

The base-to-emitter voltage drop is related to the emitter current by the expression: $I_E = I_S \exp(qV/NKT)$, where I_S and N are constants determined by the device type and $q/KT = 40 \text{ } V^{-1}$ at 25°C. Assume all diode drops and all base-emitter voltages of Q_3 are 0.7 V dc.

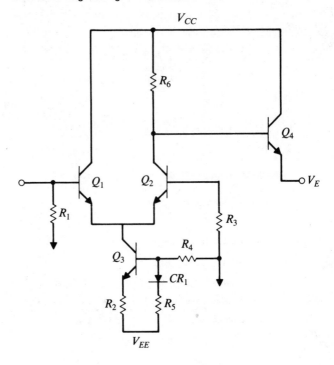

$$\beta = 200 \text{ (all transistors)}$$
$$V_{CR_1} = .7 \text{ V dc (drop accross } CR_1)$$
$$V_{CC} = 15 \text{ V dc}$$
$$V_{EE} = -15 \text{ V dc}$$
$$R_1 = 10 \text{ k}\Omega$$
$$R_2 = 5 \text{ k}\Omega$$
$$R_3 = 10 \text{ k}\Omega$$
$$R_4 = 4.3 \text{ k}\Omega$$
$$R_5 = 10 \text{ k}\Omega$$
$$R_6 = 12 \text{ k}\Omega$$

Fig. 1-53   Discrete operational amplifier front end.

SOLUTION: If the transistors are identically matched, $I_{S1} = I_{S2}$, and assume $N_1 = N_2 = 1$, because N normally varies between 1 and 2. Define V_1 and V_2 as the V_{BE} voltages of Q_1 and Q_2, respectively.

$$I_{E2} = I_{S2} \exp(40 \ V_2)$$
$$V_1 = 0.025 \ \ln(I_{E1}/I_{S1})$$
$$V_2 = 0.025 \ \ln(I_{E2}/I_{S2})$$

and the differential offset voltage ΔV is $(V_1 - V_2)$:

$$\Delta V = V_1 - V_2 = 0.025 \ \ln(I_{E1}/I_{S1}) - 0.025 \ \ln(I_{E2}/I_{S2})$$
$$\Delta V = 0.025 \ \ln[(I_{E1}/I_{S1})(I_{S2}/I_{E2})]$$

but $I_{S1} = I_{S2}$, and

$$\Delta V = 0.025 \ \ln(I_{E1}/I_{E2})$$

Now, if we can calculate the ratio of I_{E1} and I_{E2}, we can find the differential offset. Refer to Fig. 1-54. First, we must calculate I_x, the current source determined by the Q_3 circuit configuration. Neglect the base current into Q_3 because the β is so high.

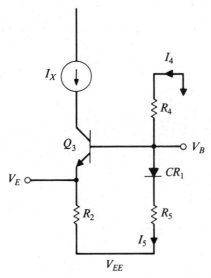

Fig. 1-54 Circuit for calculation of I_X.

$$I_4 = I_5 = \frac{0V - V_{EE} - V_{CR1}}{R_4 + R_5} = \frac{(0 + 15 - 0.7) \text{ V}}{4.3 \text{ k}\Omega + 10 \text{ k}\Omega} = 1 \text{ mA}$$

$$V_B = V_{EE} + I_5 R_5 + V_{CR1} = -15 + (10^{-3})(10^4) + 0.7 = -4.3 \text{ V dc}$$

$$V_E = V_B - V_{BE} = -4.3 - 0.7 = -5 \text{ V dc}$$

$$V_{R2} = V_E - V_{EE} = -5 + 15 = 10 \text{ V dc (voltage across } R_2)$$

Since β is large, $I_x = I_{E3} = I_{R2}$; $I_x = \beta I_{E3}/(\beta + 1)$:

$$I_{R2} = V_{R2}/R_2 = 10 \text{ V}/5 \text{ k}\Omega = 2 \text{ mA}$$

What is I_{E2} at $V_E = 2$ V dc? Label the collector of Q_2, V_{C2}.

$$I_{E2} = I_{C2} = \frac{V_{CC} - V_{C2}}{R_6} = \frac{V_{CC} - V_E - V_{BE4}}{R_6}$$

$$I_{E2} = \frac{15 - 2 - 0.7 \text{ V}}{12 \text{ k}\Omega} = 1.025 \text{ mA}$$

therefore, since $I_{E1} + I_{E2} = I_x$,

$$I_{E1} = 2 \text{ mA} - 1.025 \text{ mA} = 0.975 \text{ mA}$$

and the initial offset is

$$\Delta V_1 = 0.025 \ln(0.975/1.025) = -1.3 \text{ mV}$$

Calculate ΔV_2 for the case when $V_E = 6$ V dc

$$I_{E2} = \frac{15 - 6 + 0.7}{R_6} = 0.692 \text{ mA}$$

$$I_{E1} = 2 \text{ mA} - 0.692 \text{ mA} = 1.308 \text{ mA}$$

and

$$\Delta V_2 = 0.025 \ln(1.308/0.692) = 15.9 \text{ mV}$$

and the difference is $\Delta V_1 - \Delta V_2 = -1.3 - 15.9 = -17.2$ mV

CHECK THE SOLUTION: Check the derivation and recalculate the numerical portion of the problem.

PROBLEM 28 _____

GIVEN: The feedback amplifier shown in Fig. 1-55 with the parameters shown.

FIND: Using approximations, find the closed-loop gain e_o/e_{in}.

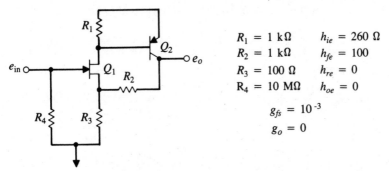

$$R_1 = 1 \text{ k}\Omega \qquad h_{ie} = 260 \text{ } \Omega$$
$$R_2 = 1 \text{ k}\Omega \qquad h_{fe} = 100$$
$$R_3 = 100 \text{ } \Omega \qquad h_{re} = 0$$
$$R_4 = 10 \text{ M}\Omega \qquad h_{oe} = 0$$

$$g_{fs} = 10^{-3}$$
$$g_o = 0$$

Fig. 1-55   Feedback amplifier for Prob. 28.

SOLUTION: First, solve for the open-loop gain. Then use Eq. (C1-2) to solve for the closed-loop gain. The open-loop ac equivalent circuit is shown in Fig. 1-56. We have a common-source JFET stage followed by a common-emitter bipolar stage. The defining equations for these two stages are given in Appendices B1 and A2, respectively. R_x of Fig. 1-56 is equal to the total loading seen by R_2 looking back into R_3 and the source of Q_1. R_x and g_o of Eq. (B2-3) are given as zero and the input

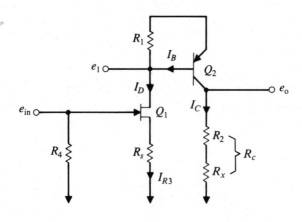

$$R_S = 90.9 \text{ } \Omega \qquad R_x = 90.9 \text{ } \Omega \qquad R_C = 1090.9 \text{ } \Omega$$

Fig. 1-56   AC equivalent circuit for the circuit for Prob. 28.

impedance at the source of Q_1 can be determined from (B2-2) to be

$$Z_{in} = \frac{1}{g_o + g_{fs}} + \frac{(R_x + R_D)g_o}{g_o + g_{fs}} + \frac{g_{fs}R_x}{g_o + g_{fs}} = \frac{1}{0 + g_{fs}} + 0 + 0 \approx \frac{1}{g_{fs}}$$

and the value of R_x can be determined as

$$R_x = R_3 \| Z_{in} = R_3 \| (1/g_{fs}) = 100 \| 10^3 = 90.9 \ \Omega$$

R_C, the open-loop collector load of Q_2, is $R_2 + R_X = 1090.9$. Now solve for R_s. Assume e_o is ac ground. The ac resistance seen looking from the source of Q_1 is $R_3 \| R_2 = 100 \| 1 \text{ k}\Omega = 90.9 \ \Omega$. Also, the input impedance looking into the base of Q_2 is $Z_{in2} \approx h_{ie} = 260 \ \Omega$ (A2-2). The equivalent drain load of Q_1 is $Z_{in2} \| R_1 = 260 \| 1000 \approx 206 \ \Omega$, and $h_{ib} \approx h_{ie}/h_{fe} = 2.6 \ \Omega$ (A2-12). Now we can solve for our open-loop gain. Let A_{v1} represent the first stage gain, A_{v2} the second-stage gain, and A_o the total open-loop gain.

$$A_{v1} = \frac{e_1}{e_{in}} = -\frac{g_{fs}R_o}{1 + g_{fs}R_s} = -\frac{(10^{-3})(206)}{1 + (10^{-3})(90.9)} = -0.189 \quad \text{(B1-5)}$$

$$A_{v2} = \frac{e_0}{e_1} \approx -\frac{R_C}{h_{ib}} = -\frac{1090.9}{2.6} = -419.6 \quad \text{(A2-1)}$$

$$A_o = A_{v1}A_{v2} = (-0.189)(-419.6) = 79.3$$

From Appendix C1 and the circuit of Fig. 1-57, we can calculate the closed-loop gain A_v. For an approximate solution, we assume for the circuit of Fig. C1-1 that $R_o = 0$, $R_{in} = \infty$, and $Z_s = 0$. Equation (C1-2) reduces to

$$A_v = \frac{A_o(R_2 + R_3)}{R_2 + R_3 + A_oR_2} = \frac{79.3(1100)}{1100 + 79.3(100)} = 9.67$$

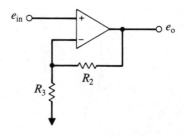

Fig. 1-57 Closed-loop simplified configuration for Prob. 28.

CHECK THE SOLUTION: Assume a change in collector current, $\Delta I_c = 1$ mA.

$$\Delta V_{R2} = \Delta I_c R_2 = (10^{-3})(10^3) = 1 \text{ V}$$

$$\Delta I_B \approx \frac{\Delta I_c}{h_{fe}} = \frac{10^{-3}}{100} = 0.01 \text{ mA}$$

$$\Delta V_B = \Delta I_B h_{ie} = (10^{-5})(260) = 2.6 \text{ mV}$$

$$\Delta I_{R1} = \frac{\Delta V_B}{R_1} = \frac{2.6 \times 10^{-3}}{10^3} = 2.6 \ \mu\text{A}$$

Now,

$$\Delta I_D = \Delta I_B + \Delta I_{R1} = 10 \ \mu\text{A} + 2.6 \ \mu\text{A} = 12.6 \ \mu\text{A}$$

Since $g_{fs} = \Delta I_D / \Delta V_{GS}$:

$$\Delta V_{GS} = \frac{\Delta I_D}{g_{fs}} = 12.6 \text{ mV}$$

$$\Delta V_{R3} = (\Delta I_D + \Delta I_C)R_3 = (1.0126)(10^{-3})(100) = 0.10126$$

$$\Delta e_{in} = \Delta V_{R3} + \Delta V_{GS} = 0.10126 + 0.0126 = 0.1139$$

$$\Delta e_o = \Delta V_{R3} + \Delta V_{R2} = 0.10126 + 1 = 1.10126$$

The closed-loop gain is

$$\frac{\Delta e_o}{\Delta e_{in}} = \frac{1.10126}{0.1139} = 9.67$$

which is the same magnitude as our solution.

PROBLEM 29

GIVEN: The temperature-dependent circuit of Fig. 1-58 is used to provide a linear output voltage as a function of temperature.

FIND: The output voltage over the temperature range of $-35°C$ to $85°C$ if the following data are supplied:

$V_{CR1} = 0.7$ V at 25°C

$TC = -2$ mV/°C (temperature coefficient of diode)

$VR_1 = 6.2$ V

SOLUTION: Assume an ideal operation amplifier. First calculate the output voltage at a room ambient of $+25°C$. We can use the following

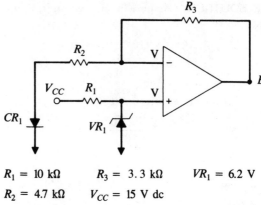

$$R_1 = 10 \text{ k}\Omega \qquad R_3 = 3.3 \text{ k}\Omega \qquad VR_1 = 6.2 \text{ V}$$
$$R_2 = 4.7 \text{ k}\Omega \qquad V_{CC} = 15 \text{ V dc}$$

Fig. 1-58   Temperature-dependent circuit of Prob. 29.

equation from Appendix C which refers to the basic op-amp circuit of Fig. 1-58.

$$E_o = \frac{(V_1 - V_2)(R_3 + R_4)R_2}{(R_1 + R_2)R_3} + \frac{(R_3 + R_4)V_2}{R_3} - \frac{V_3 R_4}{R_3} \quad \text{(C3-1)}$$

$$E_o = \frac{6.2 \text{ V}(8 \text{ k}\Omega)}{4.7 \text{ k}\Omega} - \frac{0.7 \text{ V}(3.3 \text{ k}\Omega)}{4.7 \text{ k}\Omega}$$

$$= 10.55 \text{ V} - 0.49 \text{ V} = 10.06 \text{ V dc}$$

At $+85°C$, V_{CR1} is:

$$V_{CR1} = 0.7 \text{ V} + (-2 \text{ mV/}°C)(85 - 25)°C$$

$$= 0.7 \text{ V} - 0.12 \text{ V} = 0.58 \text{ V dc}$$

E_o at $85°C$ is E_{o1}:

$$E_{o1} = 10.55 \text{ V} - \frac{0.58 \text{ V}(3.3 \text{ k}\Omega)}{4.7 \text{ k}\Omega} = 10.55 \text{ V} - 0.41 \text{ V} = 10.14 \text{ V}$$

At $-35°C$, V_{CR1} is:

$$V_{CR1} = 0.7 \text{ V} + (-2 \text{ mV/}°C)(-35 - 25)°C$$

$$= 0.7 \text{ V} + 0.12 \text{ V} = 0.82 \text{ V}$$

E_o at $-35°C$ is E_{o2}:

$$E_{o2} = 10.55 \text{ V} - \frac{0.82 \text{ V}(3.3 \text{ k}\Omega)}{4.7 \text{ k}\Omega} = 10.55 \text{ V} - 0.58 \text{ V} = 9.97 \text{ V dc}$$

The graph of E_o with temperature is shown in Fig. 1-59.

CHECK THE SOLUTION: At 25°C, $V^- = V^+ = 6.2$ V.

$$I_3 = \frac{(V^- - V_{CR1})}{R_2} = \frac{(6.2 - 0.7)\text{ V}}{4.7\text{ k}\Omega} = 1.17\text{ mA}$$

Since $I_4 = I_3$,

$$E_o = V^- + I_3R_3 = 6.2\text{ V} + (1.17\text{ mA})(3.3\text{ k}\Omega) = 10.06\text{ V dc}$$

At +85°C,

$$I_{R2} = \frac{(6.2 - 0.58)\text{ V}}{4.7\text{ k}\Omega} = 1.196\text{ mA}$$

and

$$E_{o1} = 6.2\text{ V} + (1.196\text{ mA})(3.3\text{ k}\Omega) = 10.147\text{ V dc}$$

At − 35°C,

$$I_3 = \frac{(6.2 - 0.82)\text{ V}}{4.7\text{ k}\Omega} = 1.145\text{ mA}$$

and

$$E_{o2} = 6.2\text{ V} + (1.145\text{ mA})(3.3\text{ k}\Omega) = 9.977\text{ V dc}$$

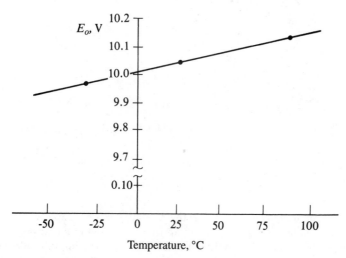

Fig. 1-59 E_o vs. temperature for the circuit of Prob. 29.

Note: The other component changes were not taken into account because they were not given. The resistor contributions can be made insignificant because they can be purchased in the range of ±0.01%.

PROBLEM 30

GIVEN: The multivibrator circuit shown in Fig. 1-60. V_x is a variable voltage used to control the frequency.

FIND:

a.   Derive an expression which describes the frequency of oscillation.

b.   What is the range of frequencies for V_x variable from +10 to +40 V dc.

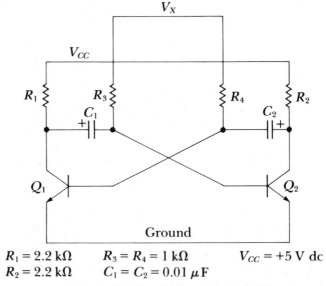

$R_1 = 2.2\ \text{k}\Omega$ $R_3 = R_4 = 1\ \text{k}\Omega$ $V_{CC} = +5\ \text{V dc}$
$R_2 = 2.2\ \text{k}\Omega$ $C_1 = C_2 = 0.01\ \mu\text{F}$

Fig. 1-60   Astable multivibrator for Prob. 30.

SOLUTION:

a.   The analysis assumes that the astable is in one stable state and then switches. Assume Q_1 is off and Q_2 is on. Assume that the base of Q_1 rises to a value to turn it on. When Q_1 turns on, the collector drops from +5 V dc to the saturation voltage of Q_1. (Assume this to be 0.2 V dc.) The (−) side of C_1 starts out at $V_{BE2(on)}$ which is assumed to be +0.7 V dc and drops 4.8 V (+5 V dc − 0.2 V dc) since the voltage across C_1 cannot change instantaneously. Thus C_1 starts charging through R_3 from − 4.1 V dc, its initial value, to a final value of + 0.7 V dc as shown in Fig. 1-61.

The equation for the voltage across the capacitor when a step function is applied to an RC circuit is given by equation (D2-10), and the derivation can be used here:

$$v_o(t) = V_F + (V_I - V_F)e^{-t/RC} \qquad \text{(D2-10)}$$

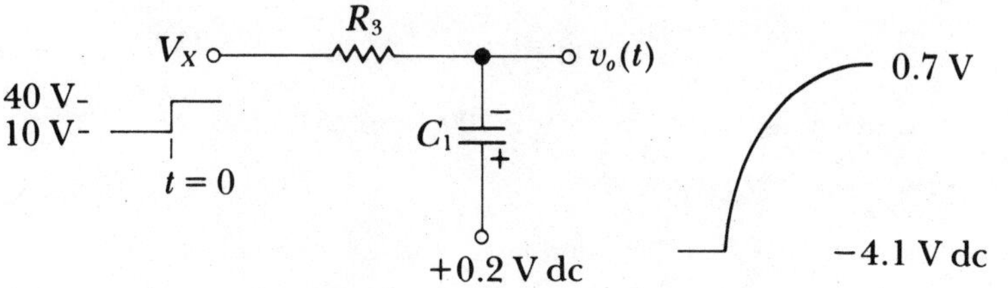

Fig. 1-61   Charging circuit for Prob. 30.

Before we go any further, let's check this equation using $(V_F = +0.7)$.

At $t = 0$: $v_o(t) = 0.7 - 4.8e^0 = -4.1$

At $t = \infty$: $v_o(t) = 0.7 - e^{-\infty} = 0.7$

This checks. Since we know the expression for $v_o(t)$, we can solve for t:

$$e^{-t/RC} = \frac{v_o(t) - V_F}{V_I - V_F}$$

$$-t/RC = \ln\left[\frac{v_o(t) - V_F}{V_I - V_F}\right]$$

$$t = (RC)\ln\left[\frac{V_I - V_F}{v_o(t) - V_F}\right] \quad \text{and} \quad f = \frac{1}{2t}$$

b.   Now let's use $V_I = -4.1$ and $v_o(t) = 0.7$, with $V_F(V_x)$ varying from +10 to +40 V dc as asked for in the problem.

At $V_F = +10$ V dc

$$t_1 = (10^3)(10^{-8})\ln\left(\frac{-4.1 - 10}{0.7 - 10}\right) = 10^{-5}\ln(1.516)$$

$$= 4.16 \, \mu s$$

Since this time is only one-half the total cycle, the frequency is:

$$f_1 = \frac{1}{2t_1} = 120.19 \text{ kHz}$$

Solving for $V_x = +40$ V dc, we find

$$t_2 = (10^{-5}) \ln\left(\frac{-4.1 - 40}{0.7 - 40}\right)$$

$$= (10^{-5}) \ln(1.122)$$

$$= 1.151 \ \mu s$$

and

$$f_2 = \frac{1}{2t} = \frac{1}{2(1.151 \times 10^{-6})} = 434.3 \text{ kHz}$$

and the range of frequencies is 120.19 to 434.3 kHz.

CHECK THE SOLUTION:

At $t_1 = 4.16 \ \mu s$, $V_F = 10$ V, and $v_o(t) = 0.7$ V,

$$v_o(t) = 10 + (-4.1 - 10) \exp\left(\frac{4.16 \times 10^{-6}}{10^{-5}}\right)$$

$$= 10 - 14.1(0.6597)$$

$$= 0.699 \text{ V} \quad \text{(checks)}$$

At $t_2 = 1.51 \ \mu s$, $V_F = 40$, $v_o(t) = 0.7$ V,

$$v_o(t) = 40 + (-4.1 - 40) \exp\left(\frac{-1.151 \times 10^{-6}}{10^{-5}}\right)$$

$$= 40 - 44.1(0.891)$$

$$= 0.699 \text{ V} \quad \text{(checks)}$$

2

CIRCUIT
ANALYSIS PROBLEMS

Almost all problems contain circuit analysis to some degree, but the problems of Chap. 2 are specific in the treatment of ac, dc, and transient circuits. Basic circuits such as resistor circuits with dc voltages and less complex ac circuits have been minimized because most recent examinations have not used these more simple circuits. However, it is important that solution techniques for such circuits be practiced because of their involvement in more complex solutions of problems. Appendices D, E, and H to K will be useful in the solution of the problem types of Chap. 2.

Equations are represented in the text as follows: (F1-4) means the equation is taken from Appendix F1 and is equation 4.

PROBLEM 1

GIVEN: The circuit of Fig. 2-1.

FIND: Assuming all transients are over and a steady state condition exists: (a) How much power is the battery delivering to the circuit? (b) How much energy is stored in the capacitor.

SOLUTION:

a. The capacitor looks like an open circuit to dc. Therefore the dc load on the battery is R_1 in series with R_2.

77

$$P = \frac{V_B^2}{R_1 + R_2} = \frac{(20)^2}{15 + 22} = 10.81 \text{ W}$$

b. The energy stored in the capacitor is a function of the dc voltage across it. This voltage is

$$V_C = \frac{R_2(V_B)}{R_1 + R_2} = 11.89 \text{ V dc}$$

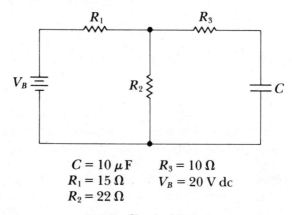

$$C = 10 \ \mu\text{F} \qquad R_3 = 10 \ \Omega$$
$$R_1 = 15 \ \Omega \qquad V_B = 20 \text{ V dc}$$
$$R_2 = 22 \ \Omega$$

Fig. 2-1   Circuit of Prob. 1.

The energy stored by the capacitor is

$$E = \frac{CV_C^2}{2} = \frac{(10^{-5})(11.89)^2}{2} = 0.707 \text{ mJ}$$

where V_C = voltage across the capacitor.

CHECK THE SOLUTION: Check each step numerically.

PROBLEM 2 _____

GIVEN: A current pulse is applied to a capacitor as shown in Fig. 2-2.

FIND: How much energy is transferred to the capacitor?

SOLUTION: Find the final voltage on the capacitor. Then the energy can be found.

$$V = \frac{1}{C} \int_0^{10^{-6}} i\, dt = \frac{it}{C} \bigg|_0^{10^{-6}} = \frac{(10)(10^{-6})}{10^{-7}} = 100 \text{ V}$$

$$E = \frac{CV^2}{2} = \frac{10^{-7}(10^4)}{2} = 0.5 \text{ mJ}$$

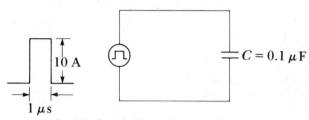

Fig. 2-2 Input pulse and circuit of Prob. 2.

CHECK THE SOLUTION: As a check, solve another way. The energy can be found by integrating the following expression (see Fig. 2-3):

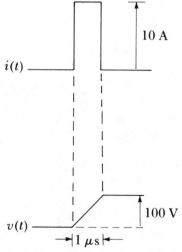

Fig. 2-3   Alternate solution waveform for Prob. 2.

$$E = \int_0^{10^{-6}} v(t)i(t)\, dt$$

where

$$V(t) = \frac{(100\ \text{V})(t)}{10^{-6}\ \text{s}} = 10^8 t$$

$$E = \int_0^{10^{-6}} (10^8)(10)t\, dt$$

$$E = \frac{10^9 t^2}{2}\Big|_0^{10^{-6}} = \frac{0.001}{2} = 0.5\ \text{mJ}$$

PROBLEM 3

GIVEN: The square wave of Fig. 2-4 is impressed upon the circuit of Fig. 2-5 at $t = 0$.

FIND:

    a. What is the output voltage for the first full cycle?

    b. Find $e_o(s)$ for the complete response to the entire periodic wave at any time t.

    c. $F(s) = \dfrac{5s + 12}{s^2 + 5s + 6}$,   find: $\mathscr{L}^{-1} F(s)$

SOLUTION:

    a. $e_i(t) = 10u(t) - 20u(t - 10^{-6}) + 20u(t - 2 \times 10^{-6}) + \cdots$

$$= \frac{10}{s} - \frac{20e^{-as}}{s} \qquad \text{for first cycle}$$

$$a = 10^{-6}$$

The transfer function for the circuit is

$$G(s) = \frac{R}{1/Cs + R} = \frac{RCs}{RCs + 1} = \frac{s}{s + 1/RC} = \frac{s}{s + 10^6}$$

$$e_o(s) = e_i(s)G(s) = \frac{10}{s + 10^6} - \frac{20e^{-as}}{s + 10^6} \qquad \text{for the first cycle}$$

From a table of Laplace transforms,

$$e_o(t) = e^{-10^6 t}[10u(t) - 20u(t - a)]$$
$$= e^{-10^6 t}[10u(t) - 20u(t - 10^6)] \qquad \text{for the first cycle}$$

    b. $e_o(s) = \dfrac{20}{s} \tanh\left(\dfrac{10^{-6}s}{2}\right)$ (E1-11)

c. By partial fraction expansion, we can determine $f(t)$ as follows:

$$F(s) = \frac{5s + 12}{s^2 + 5s + 6} = \frac{A}{s + 3} + \frac{B}{s + 2}$$

$$A(s + 2) + B(s + 3) = 5s + 12$$

At $s = -2$: $0 + B(-2 + 3) = 5(-2) + 12$

$$B = 2$$

At $s = -3$: $A(-3 + 2) + 0 = -15 + 12$

$$A = 3$$

Therefore

$$F(S) = \frac{3}{s + 3} + \frac{2}{s + 2}$$

From a table of Laplace transforms,

$$f(t) = 3e^{-3t} + 2e^{-2t}$$

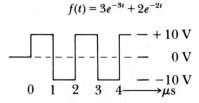

Fig. 2-4   Input waveform for Prob. 3.

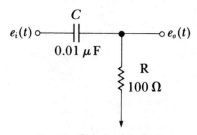

Fig. 2-5   Circuit of Prob. 3.

CHECK THE SOLUTION: Check each step for numerical accuracy and derivation accuracy.

PROBLEM 4

GIVEN: The following function:

$$G(s) = \frac{s^2 + 2s + 1}{s^3 + 6s^2 + 9s + 4} = \frac{e_o(s)}{e_i(s)}$$

FIND: It is desired to synthesize, as simply as possible, an active or passive network having this transfer function. Find a network that meets these requirements.

SOLUTION: One should always look for a factor when working a problem such as this. We can see that $(s + 1)$ is a factor for the numerator. Is it a factor for the denominator?

$$
\begin{array}{r}
s^2 + 5s + 4 \\
s + 1{\overline{\smash{\big)}\,}} s^3 + 6s^2 + 9s + 4 \\
\underline{s^3 + s^2} \\
5s^2 + 9s \\
\underline{5s^2 + 5s} \\
4s + 4 \\
\underline{4s + 4}
\end{array}
$$

$$
G(s) = \frac{(s + 1)(s + 1)}{(s + 1)(s^2 + 5s + 4)} = \frac{s + 1}{s^2 + 5s + 4}
$$

which shows $(s + 1)$ is a factor of the denominator. The denominator can be reduced further to:

$$
G(s) = \frac{s + 1}{(s + 1)(s + 4)} = \frac{1}{s + 4}
$$

We can define this as a ratio of two impedances:

$$
G(s) = \frac{Z_2}{Z_1 + Z_2} = \frac{1}{s + 4}
$$

Let $Z_2 = 1$. Therefore $Z_1 = s + 3$. A circuit that meets these requirements is shown in Fig. 2-6.

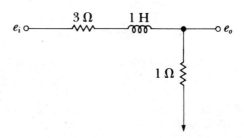

Fig. 2-6   Circuit satisfying the transfer function of Prob. 4.

CHECK:

$$G(s) = \frac{e_o}{e_i} = \frac{1}{s+3+1} = \frac{1}{s+4}$$

Note: We could have gone to Appendix K and found a circuit represented by $G(s) = A/(Ts + 1)$, since $1/(s+4)$ can be simplified to $0.25/(0.25s + 1)$. The circuit of (K1-2) from Table K1 will satisfy this problem's requirements.

PROBLEM 5

GIVEN: An input to the circuit as shown in Fig. 2-7. Assume that the diode has a forward drop of 0.7 V for all forward-current levels.

FIND: The time it takes the output to reach 3 V.

SOLUTION: The voltage across the capacitor at time zero is the forward drop of the diode, which is 0.7 V. The maximum voltage that can appear across the capacitor for large time t is

$$V_{x(max)} = \frac{R_2(V_{CC})}{R_1 + R_2} = \frac{(10 \times 10^3)(5)}{1 \times 10^3 + 10 \times 10^3} = 4.545 \text{ V}$$

because the capacitor acts like an open circuit as time grows large.

The equation which relates the output voltage $v_x(t)$ is given in Appendix D as

$$v_x(t) = V_F + (V_I - V_F)e^{-t/RC} \qquad \text{(D2-10)}$$

where V_F = final value of $v_x = v_{x(max)}$

   $V_I = 0.7$ = initial value of v_x

$v_x(t) = 4.545 + (0.7 - 4.545)e^{-t/RC}$

   $= 4.545 - 3.845e^{-t/RC}$

Check this:

At $t = 0$: $v_x(t) = 4.545 + 0.7 - 4.545 = 0.7$
At $t = \infty$: $v_x(t) = 4.545$

Therefore it checks. Now calculate the time it takes for v_x to reach 3 V.

$$v_x(t) = 3 = 4.545 - 3.845e^{-t/RC}$$
$$e^{-t/RC} = 0.4018$$
$$-t/RC = \ln 0.4018$$
$$t = 0.9118RC$$

R is determined as the Thevenin equivalent impedance looking across the capacitor. Therefore:

$$R = R_1\|R_2 = 0.9091 \text{ k}\Omega$$

and

$$t = 0.9118(0.9091 \text{ k}\Omega)(10^{-8}) = 8.289 \text{ }\mu\text{s}.$$

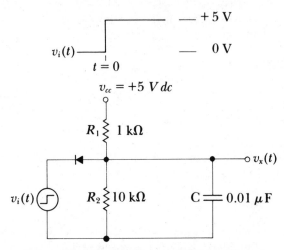

Fig. 2-7   The input waveform and circuit of Prob. 5.

CHECK THE SOLUTION: $v_x(t)$ at $t = 8.289$ μs should equal 3 V.

$$v_x(t) = 4.545 - 3.845 \exp(-8.289 \times 10^{-6}/0.909 \times 10^{-5})$$
$$= 4.545 - 1.545 = 3 \text{ V}   \text{(checks)}$$

PROBLEM 6

GIVEN: The voltage applied to a 33-pF capacitor as shown in Fig. 2-8.

FIND:

    a. Find the current waveform.

    b. What is the magnitude of current at $t = 1$ s? 5 s? 8 s?

SOLUTION:

    a. Intuitively, one can show that charge is being placed on the capacitor at a constant rate ($i = dq/dt = C\,dv/dt$) in the interval 0 to 3 s. No charge is being supplied between 3 and 6 s. Between 6 and

10 s, charge is being drawn from the capacitor at a constant rate $(i = -dq/dt = -C\,dv/dt)$. Therefore, one might expect the current to look as shown in Fig. 2-9. The current levels can be determined:

$$i_1 = C(dv/dt) = (33 \times 10^{-12})(10\text{ V})/3\text{ s} = 1.1 \times 10^{-10}\text{ A}$$

$$i_2 = -C(dv/dt) = -(33 \times 10^{-12})(10\text{ V})/4\text{ s} = -8.25 \times 10^{-11}\text{ A}$$

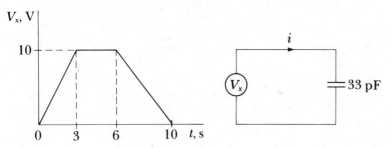

Fig. 2-8   The input voltage waveform and circuit for Prob. 6.

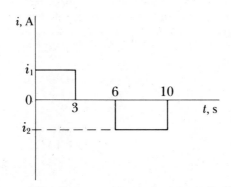

Fig. 2-9   The output current waveform for Prob. 6.

b. At $t = 1$ s: $i = 1.1 \times 10^{-10}$ A

At $t = 5$ s: $i = 0$ A

At $t = 8$ s $i = -8.25 \times 10^{-11}$

CHECK THE SOLUTION: The solution may be found mathematically. The expression defining V_x is

$$V_x(t) = \frac{(10)(t)u(t)}{3\text{ s}} - \frac{(10\text{ V})(t-3)u(t-3)}{3\text{ s}} - \frac{(10\text{ V})(t-6)u(t-6)}{4\text{ s}}$$

$$V_x(s) = \frac{10}{3s^2} - \frac{10e^{-3s}}{3s^2} - \frac{10e^{-6s}}{4s^2}$$

$$I_x(s) = \frac{V_x(s)}{Z(s)} \quad \text{and} \quad Z(s) = \frac{1}{Cs} = \frac{3.03 \times 10^{10}}{s}$$

$$= \frac{10^{-10}}{0.91s} - \frac{10^{-10}e^{-3s}}{0.91s} - \frac{10^{-10}e^{-6s}}{1.212s}$$

$$I_x(t) = 1.1 \times 10^{-10}u(t) - 1.1 \times 10^{-10}u(t-3) - 0.825 \times 10^{-10}u(t-6)$$

which checks our previous answers when evaluated at $t = 1, 5, 8$ s.

PROBLEM 7

FIND: Can $1\,\Omega$ be the characteristic impedance of a voltage source? Can $100\,\text{k}\Omega$ be the characteristic impedance of a current source? Explain.

SOLUTION:

a.   Voltage source—$1\,\Omega$ could be the characteristic impedance of a voltage source or even a current source, depending on the load. A more correct definition of a voltage source is this: a source whose output voltage is constant with changing load (because its im-

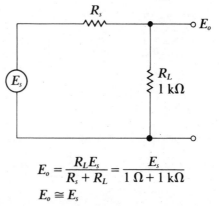

$$E_o = \frac{R_L E_s}{R_s + R_L} = \frac{E_s}{1\,\Omega + 1\,\text{k}\Omega}$$

$$E_o \cong E_s$$

Fig. 2-10   Representative voltage source circuit for Prob. 7.

pedance should be small with respect to the load). If the load is $1\,\text{k}\Omega$, $1\,\Omega$ would be a suitable impedance for a voltage source. If the load is $1\,\Omega$, it is not suitable. Refer to the circuit of Fig. 2-10.

   For $R_L = 1\,\Omega$

$$E_o = \frac{R_L E_s}{R_s + R_L} = \frac{E_s}{1+1} = 0.5 E_s$$

Note that if R_L changes from 1 kΩ to 2 kΩ in the first example, the voltage changes very little. For the second example, a 100% change in the load (from 1 to 2 Ω) yields a 16% change in the output voltage, which is surely not indicative of a voltage source.

b.   In a similar fashion, a 100-kΩ source impedance with a load of 1 Ω is certainly a current source. See Fig. 2-11.

$$I_s = \frac{E_s}{R_s + R_L} = \frac{E_s}{100 \text{ k}\Omega + 1 \text{ }\Omega}$$

$$\approx \frac{E_s}{R_s} = 10^{-5} E_s$$

If R_L changes to 2 Ω, the source current is still close to the same value.

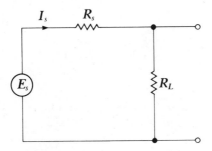

Fig. 2-11   Representative current source circuit for Prob. 7.

If $R_L = 100$ kΩ, the source is no longer a current source.

$$I_s = \frac{E_s}{R_s + R_L} = \frac{E_s}{100 \text{ k}\Omega + 100 \text{ k}\Omega} = 5 \times 10^{-6} E_s$$

If R_L changes by a factor of 2, the current source changes, an undesirable trait.

$$I_s = \frac{E_s}{100 \text{ k}\Omega + 200 \text{ k}\Omega} = 3.33 \times 10^{-6} \text{ A}$$

The change in current is 33%, and the source cannot be considered a current source.

PROBLEM 8

GIVEN: The circuit of Fig. 2-12. The switch has been closed long enough for steady-state conditions to exist. Assume that at $t = 0$, the switch is opened.

FIND: The expression for $v_o(t)$, and calculate for $t = 0.1$ s.

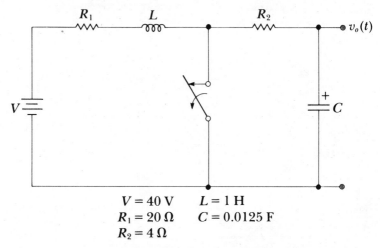

$$V = 40 \text{ V} \qquad L = 1 \text{ H}$$
$$R_1 = 20 \ \Omega \qquad C = 0.0125 \text{ F}$$
$$R_2 = 4 \ \Omega$$

Fig. 2-12   Circuit configuration for Prob. 8.

SOLUTION: Opening the switch is the equivalent of impressing a step function at time $t = 0$; therefore, $V(s) = V/s$. The initial voltage across the capacitor is zero, so $q(0^+)$ is zero. The initial current through the inductor is obtained readily from the expression

$$i_L(0^+) = V/R_1$$

The expression which describes the circuit at $t = (0^+)$ is as follows (let $R = R_1 + R_2$):

$$V(s) = (1/Cs)[I(s) + q(0^+)] + L[sI(s) - i_L(0^+)] + RI(s)$$
$$V/s = (1/Cs)I(s) + LsI(s) - VL/R_1 + RI(s)$$
$$\frac{V}{s} + \frac{VL}{R_1} = I(s)\left(\frac{1 + RCs + LCs^2}{Cs}\right)$$

and

$$I(s) = \frac{V(Ls + R_1)}{R_1 L[s^2 + (R/L)s + (1/LC)]}$$

$$v_o(s) = I(s)(1/Cs)$$

$$= \frac{V(Ls + R_1)}{R_1 L Cs[s^2 + (R/L)s + (1/LC)]}$$

Substitute numerical values:

$$v_o(s) = \frac{40(s + 20)}{(20)(1)(0.0125)s(s^2 + 24s + 80)}$$

$$= \frac{160(s + 20)}{s(s^2 + 24s + 80)}$$

Normally, when working expressions in polynomials of s, it is good practice to try to find a factor that is common to both the numerator and denominator so that the expression can be simplified. Such is the case with this problem.

$$v_o(s) = \frac{160(s + 20)}{s(s + 20)(s + 4)} = \frac{160}{s(s + 4)}$$

From a table of Laplace transforms,

$$v_o(t) = 40(1 - e^{-4t})$$

At $t = 0.1$ s,

$$v_o(t) = 40(1 - e^{-0.4}) = 13.19 \text{ V}$$

CHECK THE SOLUTION: As t becomes large, C becomes an open circuit and $v_o = V$.

$$v_o(t) = 40(1 - e^{-\infty}) = 40$$

When the switch is first opened the capacitor looks like a short circuit and the current flowing through L is that flowing through R_1, R_2, and C. At $t = 0$, $i = V/R_1 = 2$A and $V_o(t) = 40(1 - e^0) = 0$. Also, t at $V_o = 13.19$ V,

$$e^{-4t} = \frac{1 - v_o(t)}{V} = 0.67 \quad \text{and} \quad t = 0.1 \text{ s}$$

PROBLEM 9

GIVEN: The circuit of Fig. 2-13. The switch is closed at $t = 0$. Neglect the effects of R in determining the transient state.

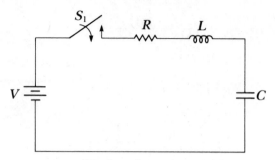

Fig. 2-13   Circuit configuration for Prob. 9.

FIND: Determine the maximum voltage that appears across the capacitor if $V = 100$ V dc.

SOLUTION: There are no initial charges or currents to consider. Write the expression for $Z(s)$ and $I(s)$, and for $V_C(s)$, the voltage across the capacitor.

$$Z(s) = \frac{1}{Cs} + Ls$$

$$= \frac{1 + LCs^2}{Cs}$$

$$V(s) = \frac{V}{s}$$

$$I(s) = \frac{V(s)}{Z(s)} = \frac{VC}{1 + LCs^2} = \frac{V}{L[s^2 + (1/LC)]}$$

$$V_C(s) = I(s)(1/Cs) = \frac{V}{LCs[s^2 + (1/LC)]}$$

From a table of Laplace transforms:

$$V_C(t) = \frac{V\{1 - \cos[t/(LC)^{1/2}]\}}{LC(1/LC)} = V\{1 - \cos[t/(LC)^{1/2}]\}$$

$$\frac{dV_C(t)}{dt} = \frac{V\sin[t/(LC)^{1/2}]}{(LC)^{1/2}}$$

Set the derivative equal to 0. We find

$$\sin[t/(LC)^{1/2}] = 0$$

or
$$\frac{t}{(LC)^{1/2}} = \pi$$

$$t = \pi(LC)^{1/2}$$

and $V_C(t) = V(1 - \cos \pi) = V(1 + 1) = 2V = 200 \text{ volts}$

CHECK THE SOLUTION: It is apparent that $t = \pi(LC)^{1/2}$ is the maximum value. Let's check the solution for $V_C(t)$, however. We know that the voltage across the capacitor is 0 at $t = 0^+$ because there is no current flowing and the capacitor is a short circuit:

At $t = 0$: $V_C(t) = V(1 - 1) = 0$ (checks)

PROBLEM 10

GIVEN: The circuit of Fig. 2-14 with the values as indicated. Z_s is the fixed internal impedance of the source. Maximum power is to be delivered to the load.

FIND:

    a.   The turns ratio of T_1

    b.   The power supplied to the load

SOLUTION:

    a.   For maximum power transfer, the impedance seen by the generator should be equal in magnitude to that of its source. Find the magnitude of $Z(s)$ first.

$$Z(j\omega) = 200 + j100 = 223.6\underline{/26.56°}$$
$$|Z| = 223.6$$

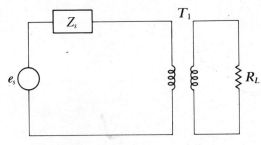

$Z_s = 200 + j100 \ \Omega$ $e_s = 141.4 \sin \omega t$
$R_L = 10 \ \Omega$

Fig. 2-14   Circuit configuration for Prob. 10.

Therefore $Z_p = 223.6$, the primary impedance of the transformer. The turns ratio n relates the primary and secondary resistances as follows (see Appendix D3):

$$n = N_p/N_s \qquad n^2 = Z_p/Z_s$$
$$n^2 = 223.6/10 = 22.36$$
$$n = (22.36)^{1/2} = 4.73$$

b. Let Z_T represent the total impedance seen by the source, I_T represent the primary current, and P_T represent the total power delivered to the transformer primary. Since we assume the transformer to be ideal, the power delivered to the load is the same as that delivered to the primary.

$$Z_T = 200 + 223.6 + j100$$
$$= 423.6 + j100 = 434.7\underline{/13.3°}$$

$$I_T = \frac{100}{435.2\underline{/13.3°}} = 0.23\underline{/-13.3°} \qquad (\text{rms} = .707\,V_p)$$

$$P_T = I_T^2 Z_p = (0.23)^2(223.6) = 11.828 \text{ W}$$

CHECK THE SOLUTION: Check the power delivered when $Z_p = 220$ and 225 Ω. When $Z_p = 220$ Ω, $Z_T = 431.74\underline{/13.39°}$, $I_T = 0.2316\underline{/-13.39°}$, and $P_T = 11.8$ W. For $Z_p = 225$ Ω, $Z_T = 436.6\underline{/13.24°}$, $I_T = 0.229\underline{/-13.24°}$, and $P_T = 11.8$ W. Both are less than the maximum, 11.828 W.

PROBLEM 11 _____

GIVEN: The circuit of Fig. 2-15.

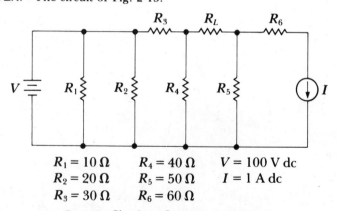

$R_1 = 10\ \Omega$	$R_4 = 40\ \Omega$	$V = 100$ V dc
$R_2 = 20\ \Omega$	$R_5 = 50\ \Omega$	$I = 1$ A dc
$R_3 = 30\ \Omega$	$R_6 = 60\ \Omega$	

Fig. 2-15   Circuit configuration for Prob. 11.

FIND: The current through R_L for values of $R_L = 10, 20, 30, 40,$ and $50\ \Omega$.

SOLUTION: This is a good problem because it makes us review several circuit analysis techniques based on superposition, voltage sources, current sources, and Thevenin's technique. Thevenin's technique is hinted at when the problem lets the load assume many values. The only efficient means of solving the problem is to use Thevenin's or Norton's theorem. R_1 and R_2 do not enter into the problem because we have to assume V is an ideal voltage source with zero internal resistance, and no matter what load we put across V, it will not change the voltage. Similarly, we can neglect R_6 because no matter what we put in series with a current source we will not change the current, since I is assumed to be ideal with infinite source impedance. Now our problem becomes that as shown in Fig. 2-16. Let's use superposition to find V_{TH}, the voltage across R_L. We must remove R_L.

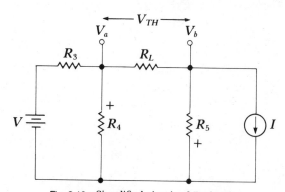

Fig. 2-16   Simplified circuit of Prob. 11.

$$V_a = \frac{VR_4}{R_3 + R_4}$$

$$= \frac{(100)(40)}{30 + 40} = 57.14 \text{ V dc}$$

$$V_b = -IR_5 = -(1)(50) = -50 \text{ V dc}$$

$$V_{TH} = V_a - V_b = 57.14 - (-50) = 107.14 \text{ V dc}$$

Find Z_{TH}: Looking back into the R_L terminal we find that Z_{TH} is $R_3 \| R_4 + R_5$. (Open-circuit the current source and short-circuit the voltage source.)

$$Z_{TH} = R_3 \| R_4 + R_5 = 30 \| 40 + 50 = 67.14\ \Omega$$

Our circuit now becomes that of Fig. 2-17:

$$I_L = \frac{V_{TH}}{R_L + Z_{TH}}$$

At $R_L = 10$: $I_L = \dfrac{107.14}{10 + 67.14} = 1.3889$ A dc

At $R_L = 20$: $I_L = \dfrac{107.14}{20 + 67.14} = 1.2295$ A dc

At $R_L = 30$: $I_L = \dfrac{107.14}{30 + 67.14} = 1.1029$ A dc

At $R_L = 40$: $I_L = \dfrac{107.14}{40 + 67.14} = 1.0000$ A dc

At $R_L = 50$: $I_L = \dfrac{107.14}{50 + 67.14} = 0.9146$ A dc

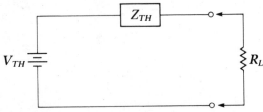

Fig. 2-17   Thevenin's equivalent for Prob. 11.

CHECK THE SOLUTION: Let's check each step numerically on the examination; but let's choose $R_L = 20$ as a check on the validity of our derivation. Choose the currents as shown in Fig. 2-18. At $R = 20\ \Omega$,

$I_{R_L} = 1.2295$ A dc

$I_{R5} = I_{R_L} - I$

$\quad = 1.2295 - 1 = 0.2295$ A dc

$V_b = I_{R5}R_5 = (0.2295)(50) = 11.475$ V dc

$V_a = V_b + I_{R_L}R_L = 11.475 + (1.2295)(20) = 36.065$ V dc

$I_{R3} = \dfrac{V - V_a}{R_3} = \dfrac{100 - 36.065}{30} = 2.131$ A dc

$I_{R4} = \dfrac{V_a}{R_4} = \dfrac{36.065}{40} = 0.9016$ A dc

Now, I_{R3} should equal $I_{R4} + I_{R_L}$.

$$I_{R4} + I_{R_L} = 2.131 \text{ A dc}$$

which checks with the value of $I_{R3} = 2.131$ A previously obtained.

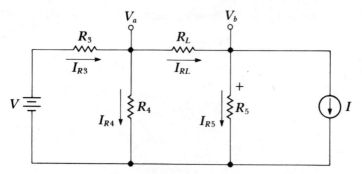

Fig. 2-18   Check circuit for Prob. 11.

PROBLEM 12

GIVEN: The circuit of Fig. 2-19. A step input is applied and it is desired that the current waveform never go negative.

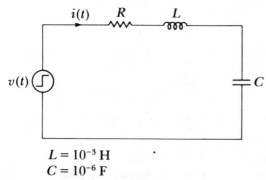

$$L = 10^{-3} \text{ H}$$
$$C = 10^{-6} \text{ F}$$

Fig. 2-19   Circuit configuration of Prob. 12.

FIND: With the values shown, what is the minimum value R can assume for this condition?

SOLUTION: Find $Z(s)$, then $I(s)$, and $i(t)$.

$$Z(s) = \frac{1}{Cs} + Ls + R$$

$$= \frac{L}{s}\left(s^2 + \frac{Rs}{L} + \frac{1}{LC}\right)$$

$$I(s) = \frac{V(s)}{Z(s)}$$

$$= \frac{V/L}{s^2 + (R/L)s + 1/LC}$$

From a table of Laplace transforms, the inverse of $I(s)$ can be found:

$$F(s) = \frac{1}{s^2 + 2as + b^2}$$

The point we are after is that value at which the circuit is critically damped; that is, $b^2 = a^2$

$$a = \frac{R}{2L}; \quad b^2 = \frac{1}{LC}$$

$$\left(\frac{R}{2L}\right)^2 = \frac{1}{LC}$$

$$R = \left(\frac{4L}{C}\right)^{1/2} = \left(\frac{4 \times 10^{-3}}{10^{-6}}\right)^{1/2} = 63.2 \ \Omega$$

CHECK THE SOLUTION:

$$I(s) = \frac{V/L}{[s^2 + (R/L)s + (1/LC)^2]}$$

$$= \frac{V/L}{[s^2 + 63.2 \times 10^3 + 10^9]}$$

$$\left(\frac{R}{2L}\right)^2 = \frac{1}{LC}$$

$$\left(\frac{63.2 \times 10^3}{2 \times 10^{-3}}\right)^2 = 10^9$$

$$10^9 = 10^9 \qquad \text{(checks)}$$

PROBLEM 13

GIVEN: The circuit of Fig. 2-20 with an input waveform to the circuit as shown. We want an output that is 20 V peak and whose width is 10 μs wide at 4 V. R_1 is given as 100 Ω.

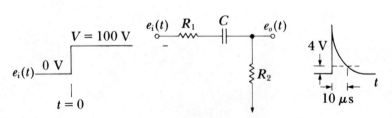

Fig. 2-20   Circuit and input/output waveforms of Prob. 13.

FIND: R_2 and C.

SOLUTION: Let's solve for $Z(s)$, $I(s)$, and then $e_o(s)$ using mathematics:

$$Z(s) = R_1 + R_2 + 1/Cs$$

$$= \frac{(R_1 + R_2)Cs + 1}{Cs}$$

$$e_i(s) = V/s$$

$$I(s) = \frac{e_i(s)}{Z(s)} = \frac{VCs}{s[(R_1 + R_2)Cs + 1]}$$

$$= \frac{VC}{(R_1 + R_2)Cs + 1}$$

$$e_o(s) = I(s)R_2 = \frac{VCR_2}{(R_1 + R_2)Cs + 1} = \frac{VR_2}{R_1 + R_2}\left[\frac{1}{s + \dfrac{1}{(R_1 + R_2)C}}\right]$$

From a table of Laplace transforms we find

$$e_o(t) = \frac{VR_2 \exp\left[\dfrac{-t}{(R_1 + R_2)C}\right]}{R_1 + R_2}$$

We know that initially the capacitor looks like a short circuit to a step input; therefore, the maximum voltage out is the ratio

$$\frac{e_o}{e_i} = \frac{R_2}{R_1 + R_2} = \frac{20\,\text{V}}{100\,\text{V}} - = 0.2$$

Solving, we obtain

$$R_2 = 25\,\Omega$$

Solve for C, substituting into our $e_o(t)$ equation the following values: $t = 10\,\mu s$; $R_1 = 100\,\Omega$; $R_2 = 25\,\Omega$; $V = 100\,\text{V}$, and $e_o(t) = 4\,\text{V}$.

$$4 = \frac{100(25)}{125} \exp\left(\frac{-10^{-5}}{125C}\right)$$

$$0.2 = \exp\left(\frac{-10^{-5}}{125C}\right)$$

$$\ln(0.2) = \frac{-10^{-5}}{125C}$$

$$\ln(5) = \frac{10^{-5}}{125C}$$

$$C = 4.97 \times 10^{-8} = 49.7\,\text{nF}$$

CHECK THE SOLUTION: Refer to the $e_o(t)$ equation. When the step is first applied ($t = 0$), we should have e_o equal to $R_2/(R_1 + R_2)$ times the input voltage. This is true:

$$e_o(t) = \frac{VR_2}{R_1 + R_2} = \frac{(100)(25)}{125} = 20 \text{ V}$$

At $t = \infty$, $e_o(t)$ should be zero, and this is also true because the exponent term becomes zero. The last check is the capacitance value check. Check $e_o(t)$ with $C = 49.7 \times 10^{-9}$:

$$e_o(t) = \frac{V(R_2)}{R_1 + R_2} \exp\left[\frac{-10^{-5}}{(125)(49.7 \times 10^{-9})}\right] = 4 \text{ V}$$

and our check is complete.

PROBLEM 14

GIVEN: The following expression for a filter:

$$G(s) = \frac{e_o(s)}{e_i(s)} = \frac{2500s}{(s + 1000)(s + 10,000)}$$

FIND:

a. The type of filter described by the equation.

b. The attenuation in decibels at frequencies of 10 Hz and 5 kHz · with respect to that of midband.

c. A passive circuit that meets the given transfer function.

SOLUTION:

a. We can draw a Bode plot as shown in Fig. 2-21. We have a zero at zero frequency and poles at $\omega = 1000$ and $\omega = 10,000$ rad/s, which corresponds to frequencies of 159 and 1590 Hz, respectively (divide ω by 2π). It is obvious from the Bode plot that we have a bandpass filter. The dashed line represents the exact response, while the solid line represents the straight-line approximation.

b. The midband frequency can be found by finding the geometric midpoint of the response. This is found by increasing the lower-pole frequency by a factor while decreasing the upper-pole frequency by the reciprocal of that factor as follows, until we get the same frequency for f_L and f_H:

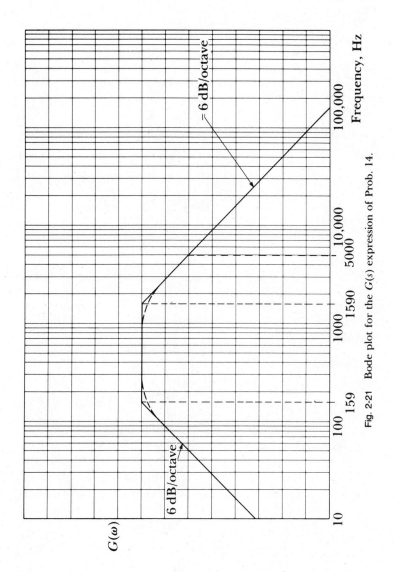

Fig. 2-21 Bode plot for the $G(s)$ expression of Prob. 14.

99

f_L	f_H	Factor
159	1590	
318	795	$(2, \frac{1}{2})$
502	503	$(1.58, 1/1.58)$

This tells us our midband frequency is 503 Hz. Calculate the gain at this frequency:

At 503 Hz: $\qquad s = j\omega = j2\pi(503) = 3160$

$$G(j3160) = \frac{2500(j3160)}{(j3160 + 1000)(j3160 + 10{,}000)}$$

$$= \frac{j2500}{(0.3165 + j)(j3160 + 10{,}000)}$$

$$= \frac{2500\underline{/90°}}{(1.05\underline{/72.44°})(10{,}487\underline{/17.54°})} \approx 0.227\underline{/0°}$$

The fact that our phase is close to 0° demonstrates that our pole at 159 Hz has cancelled the effect of our zero at zero frequency, and that our pole at 1590 has not started to attenuate our signal; thus, we are close to midband. Calculate the attenuation at 10 Hz which the problem asked for:

At 10 Hz: $\qquad s = j\omega = j62.8$

$$G(j62.8) = \frac{(2500)(j62.8)}{(1000 + j62.8)(10{,}000 + j62.8)}$$

$$= \frac{j2500}{(15.92 + j)(10{,}000 + j62.8)}$$

$$\approx \frac{2500\underline{/90°}}{(15.95\underline{/3.6°})(10{,}000\underline{/0.4°})} = 0.0157\underline{/86°}$$

The 86° is close to 90°, which approximates the pure lead network which we expected. Our attenuation at 10 Hz relative to midband is

$$\text{dB} = 20 \log (0.0157/0.227) = (20)(-1.16) = -23.2$$

What is our attenuation at 5 kHz?

At 5 kHz: $s = j\omega = 2\pi(5000) = 3.14 \times 10^4$

$$G(j31,400) = \frac{2500(j3.14 \times 10^4)}{(j3.14 \times 10^4 + 1000)(j3.14 \times 10^4 + 10^4)}$$

$$= \frac{j2500}{(0.0318 + j)(10^4 + j3.14 \times 10^4)}$$

$$\approx \frac{2500\underline{/90°}}{(1\underline{/88.2°})(3.295 \times 10^4\underline{/72.3°})} = 75.86 \times 10^{-3}\underline{/-70.5°}$$

The phase angle of −70.5° tells us our pole at 1590 Hz is starting to dominate and that we are past our 3-dB point on the roll-off. The attenuation in dB is

$$dB = 20 \log (0.07586/0.227) = 20(-0.476) = -9.52$$

A circuit that meets these requirements can be synthesized by reviewing the Bode plots of Appendix K, Table K1.

We find that the circuit represented by Eq. (K1-11) of Table K1 can be used to synthesize this function. Get the original expression in the proper format as follows (see Fig. 2-22):

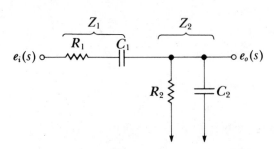

Fig. 2-22   The circuit of Eq. (K1-11) used in the solution of Prob. 14.

$$G(s) = \frac{2500s}{(s + 1000)(s + 10,000)}$$

$$= \left[\frac{2500}{(1000)(10,000)}\right]\left[\frac{s}{(10^{-3}s + 1)(10^{-4}s + 1)}\right]$$

$$= \frac{2.5 \times 10^{-4}s}{(10^{-3}s + 1)(10^{-4}s + 1)}$$

From (K1-11):

$$G(s) = \frac{T_1 s}{(T_3 s + 1)(T_4 s + 1)}$$

$$T_1 = 2.5 \times 10^{-4} \qquad T_3 = 10^{-3} \qquad T_4 = 10^{-4}$$

$$R_2 C_1 = T_1 = 2.5 \times 10^{-4}$$

$$R_1 C_1 R_2 C_2 = T_3 T_4 = 10^{-7}$$

$$T_3 + T_4 = (R_1 + R_2)C_1 + R_2 C_2$$
$$= 10^{-3} + 10^{-4} = 1.1 \times 10^{-3}$$

Let's solve for the various component values. Select one convenient component value, say R_2. Let $R_2 = 10$ kΩ. Then

$$C_1 = (2.5 \times 10^{-4}) \div (10 \times 10^3) = 2.5 \times 10^{-8}$$

Also:

$$R_1 C_2 = 10^{-7}/(2.5 \times 10^{-4}) = 4 \times 10^{-4}$$

$$C_2 = 4 \times 10^{-4}/R_1$$

$$2.5 \times 10^{-8} R_1 + 10 \times 10^3 (2.5 \times 10^{-8} + 4 \times 10^{-4}/R_1) = 1.1 \times 10^{-3}$$

$$R_1^2 - 34 \times 10^3 R_1 + 160 \times 10^6 = 0$$

$$R_1 = \{34 \times 10^3 \pm [(-34 \times 10^3)^2 - 640 \times 10^6]^{1/2}\} \left(\frac{1}{2}\right)$$

$$= 5.64 \text{ k}\Omega, 28.36 \text{ k}\Omega$$

Choose $R_1 = 28.36$ kΩ. Then:

$$C_2 = 1.41 \times 10^{-8}$$

$$R_1 C_1 R_2 C_2 = (28.36 \times 10^3)(2.5 \times 10^{-8})(10 \times 10^3)(1.41 \times 10^{-8}) = 10^{-7}$$

This checks. Also:

$$R_1 C_1 + R_2(C_1 + C_2) = (28.36 \times 10^3)(2.5 \times 10^{-8}) + 10 \times 10^3(3.91 \times 10^{-8})$$
$$= 1.1 \times 10^{-3}$$

And this checks with our original value. Our circuit becomes that of Fig. 2-23. Of course, in actual practice we would use the closest standard values with the tolerance a function of the accuracy desired.

CHECK THE SOLUTION: From (D4-3), $f_m = (f_L f_H)^{1/2} = 502.8$ Hz (vs. 503 Hz). Also,

$$G(s) = \frac{R_2C_1s}{R_1R_2C_1C_2s^2 + (R_1C_1 + R_2C_2 + R_2C_1)s + 1}$$

$$= \frac{(10^4)(2.5 \times 10^{-8})s}{10^{-7}s^2 + 1.1 \times 10^{-3}s + 1}$$

$$= \frac{2500s}{s^2 + 1.1 \times 10^4s + 10^7}$$

$$= \frac{2500s}{(s + 1000)(s + 10,000)}$$

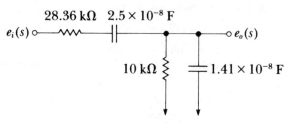

Fig. 2-23 Circuit solution to Prob. 14.

PROBLEM 15

GIVEN: The circuit of Fig. 2-24. S_1 is initially closed and S_2 is open. Assume steady-state conditions have been reached. At time $t = 0$, S_1 is opened and S_2 is closed.

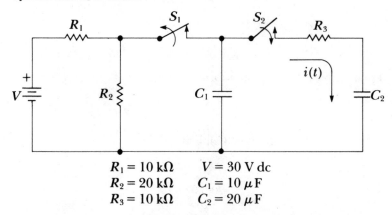

$R_1 = 10\ k\Omega$	$V = 30\ V\ dc$
$R_2 = 20\ k\Omega$	$C_1 = 10\ \mu F$
$R_3 = 10\ k\Omega$	$C_2 = 20\ \mu F$

Fig. 2-24 Circuit configuration for Prob. 15.

FIND: The steady-state voltage reached across C_2.

SOLUTION: The voltage that C_1 will charge to before S_1 is opened is

$$V_{C1} = \frac{R_2(V)}{R_1 + R_2} = \frac{20 \times 10^3(30)}{10 \times 10^3 + 20 \times 10^3} = 20 \text{ V dc}$$

The total series circuit capacitance after S_1 is opened and S_2 is closed can be found as follows:

$$C_T = \frac{C_1 C_2}{C_1 + C_2} = 6.67 \ \mu\text{F}$$

The initial charge on C_1 before S_2 is closed is

$$q_1(0^+) = C_1 V = (10^{-5})(20) = 20 \times 10^{-5} \text{ coulombs}$$

The circuit equation can now be written for the case when S_1 is opened and S_2 is closed as follows [$q_1(0^+)$ becomes a minus after S_2 is closed]:

$$1/(C_1 s)[I(s) + q_1(0^+)] + 1/(C_2 s)[I(s) + q_2(0^+)] + R_3 I(s) = 0$$
$$I(s)(1/C_1 s + 1/C_2 s + R_3) + 1/(C_1 s)(-20 \times 10^{-5}) = 0$$

Solving, we find

$$20/s = I(s)\left[\left(\frac{1}{C_1} + \frac{1}{C_2}\right)\frac{1}{s} + R_3\right]$$

$$I(s) = \frac{20}{R_3}\left[\frac{1}{s + 1/R_3 C_T}\right]$$

$$V_2(s) = I(s)(1/C_2 s)$$

$$= \frac{20}{R_3 C_2 s}\left(\frac{1}{s + 1/R_3 C_T}\right)$$

From a table of Laplace transforms we find:

$$v_2(t) = (20/R_3 C_2)\left\{R_3 C_T\left[1 - \exp\left(\frac{-t}{R_3 C_T}\right)\right]\right\}$$

$$= [(20 C_T)/C_2]\left[1 - \exp\left(\frac{-t}{R_3 C_T}\right)\right]$$

V_F occurs at $t = \infty$:

$$V_F = v_2(\infty) = \frac{20(6.67 \times 10^{-6})}{20 \times 10^{-6}} = 6.67 \text{ V}$$

which is what the problem asked for.

CHECK THE SOLUTION: The initial energy available is

$$W_I = \frac{C_1 V_{C1}^2}{2} = \frac{(10^{-5})(20)^2}{2} = 2 \text{ mJ}$$

Since $V_{C1} = V_{C2} = V_F$, the energy at $t = \infty$ is

$$W_F = \frac{C_1 V_F^2}{2} + \frac{C_2 V_F^2}{2} = 0.667 \text{ mJ}$$

The remainder of the energy to be dissipated in R_3 is

$$W_{R3} = \int_0^\infty [i(t)]^2 (R_3) \, dt = R_3 \int_0^\infty \frac{20^2}{R_3} \exp\left(\frac{-2t}{R_3 C_T}\right) dt = 1.33 \text{ mJ}$$

PROBLEM 16 _____

GIVEN: It is desired to construct a low-pass filter whose half-power point is 25 kHz using a 1000-pF capacitor and standard 1% resistor values. We would also like to construct a high-pass filter using the same capacitor and 1% resistors with a half-power frequency of 5 kHz.

FIND: Design these circuits and show the values. Can these circuits be cascaded to obtain a bandpass with half-power points at 5 kHz and 25 kHz?

SOLUTION: For our low-pass filter at 25 kHz, we need a function of the form $G(s) = 1/(\tau s + 1)$. We can find this as Eq. (K1-1) in Table K1 of Appendix K. Draw the circuit of Fig. 2-25.

$$G_1(s) = \frac{1}{T_1 s + 1}$$

$$T_1 = R_1 C = 1/(2\pi f) = 6.37 \times 10^{-6} \qquad \text{(K1-2)}$$

$R_1 = 6.37 \text{ k}\Omega$ (use nearest standard value of 6.34 kΩ from
Appendix H1)

For our high-pass of 5 kHz we will use the circuit of (K1-3), shown as Fig. 2-26.

$$G_2(s) = \frac{T_2 s}{T_2 s + 1}$$

$$T_2 = R_2 C = \frac{1}{2\pi f} = 31.83 \times 10^{-6} \qquad \text{(K1-3)}$$

$R_2 = 31.83 \text{ k}\Omega$ (use the nearest standard value of 31.6 kΩ
from Appendix H1)

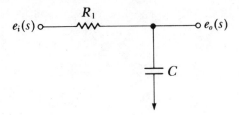

Fig. 2-25   Low-pass solution to Prob. 16.

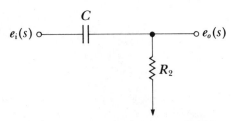

Fig. 2-26   High-pass solution to Prob. 16.

Now, can we cascade the two for a bandpass? The answer is obviously no, because one circuit loads the other and gives us a transfer function altered from that desired. Derive the expression for $e_o(s)/e_i(s) = G_3(s)$ for the circuit of Fig. 2-27 and show that $G_3(s) \neq G_1(s)G_2(s)$.

$$Z_2(s) = R_2 + 1/Cs = \frac{R_2Cs + 1}{Cs} = \frac{N_2}{D_2}$$

$$Z(s) = Z_2(s)\|(1/Cs) = \frac{N_2/Cs}{N_2 + D_2(1/Cs)}$$

$$= \frac{R_2Cs + 1}{(R_2Cs + 1 + 1)Cs} = \frac{R_2Cs + 1}{(R_2Cs + 2)Cs} = \frac{N(s)}{D(s)}$$

$$\frac{e_1(s)}{e_i(s)} = \frac{Z(s)}{R_1 + Z(s)} = \frac{N(s)}{N(s) + R_1D(s)} = \frac{R_2Cs + 1}{R_2Cs + 1 + R_1Cs(R_2Cs + 2)}$$

$$= \frac{R_2Cs + 1}{R_1R_2C^2s^2 + (R_2C + 2R_1C)s + 1}$$

$$\frac{e_o(s)}{e_1(s)} = \frac{R_2}{R_2 + 1/(Cs)} = \frac{R_2Cs}{R_2Cs + 1}$$

$$G_3(s) = \frac{e_o(s)}{e_i(s)} = \left[\frac{e_o(s)}{e_1(s)}\right]\left[\frac{e_1(s)}{e_i(s)}\right] = \frac{R_2Cs}{R_1R_2C^2s^2 + (2R_1 + R_2)Cs + 1}$$

$$G_1(s)G_2(s) = \left(\frac{1}{R_1Cs + 1}\right)\left(\frac{R_2Cs}{R_2Cs + 1}\right) = \frac{R_2Cs}{R_1R_2C^2s^2 + (R_1 + R_2)Cs + 1}$$

Note: By inspection, we see that $G_3(s) \neq G_1(s)G_2(s)$.

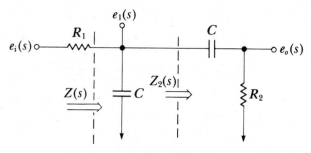

Fig. 2-27 $G_1(s)G_2(s)$ for Prob. 16.

CHECK THE SOLUTION: At 25 kHz, $X_C = R_1$; check:

$$X_C = \frac{1}{2\pi fC} = \frac{1}{2\pi(25 \times 10^3)(10^{-9})} = 6.37 \text{ k}\Omega$$

This checks. Also, at 5 kHz, $X_C = R_2$; check:

$$X_C = \frac{1}{2\pi fC} = \frac{1}{2\pi(5 \times 10^3)(10^{-9})} = 31.83 \text{ k}\Omega$$

and this checks.

PROBLEM 17

GIVEN: The filter shown in Fig. 2-28. A voltage square wave shown in Fig. 2-29 is applied to the input, and it is desired to extract the fundamental and attenuate the harmonics.

FIND: Since we can't filter perfectly, what is the percent distortion we get from the first three contributing harmonics?

SOLUTION: The transfer function that describes this circuit is given in Appendix K as (K1-9).

$$G(s) = \frac{1}{(T_3s + 1)(T_4s + 1)} = \frac{1}{T_3T_4s^2 + (T_3 + T_4)s + 1} \quad (K1-9)$$

$$T_3T_4 = R_1R_2C_1C_2 = R^2C^2 = 10^{-8}$$

$$T_3 + T_4 = R_1(C_1 + C_2) + R_2C_2 = 3RC = 3 \times 10^{-4}$$

$$G(s) = \frac{1}{10^{-8}s^2 + 3 \times 10^{-4}s + 1}$$

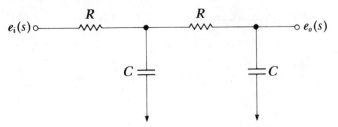

$R = 10\ k\Omega$
$C = 0.01\ \mu F$

Fig. 2-28   Filter circuit of Prob. 17.

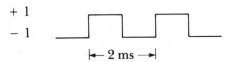

Fig. 2-29   Input voltage waveform of Prob. 17.

The fundamental frequency of the square wave is $1/T = 500$ Hz. From Appendix I, we obtain the Fourier expression for the square wave. There is no dc term because the waveform is symmetrical about the abscissa.

$$e_o(t) = \frac{4 \sin \omega t}{\pi} + \frac{4 \sin 3\omega t}{3\pi} + \frac{4 \sin 5\omega t}{5\pi} + \frac{4 \sin 7\omega t}{7\pi} + \cdots \quad \text{(I-2)}$$

Let's find the magnitude of each harmonic as it exits the filter.

$$G(j\omega) = \frac{1}{(1 - 10^{-8}\omega^2) + j3 \times 10^{-4}\omega}$$

For $f = 500$ Hz, $\omega = 3142$:

$$G(j3142) = \frac{1}{(1 - 0.099) + j0.94} = 0.768\underline{/-46.2°}$$

For $f = 1500$ Hz, $\omega = 9425$:

$$G(j9425) = \frac{1}{(1 - 0.89) + j2.83} = 0.353\underline{/-87.8°}$$

For $f = 2500$, $\omega = 1.57 \times 10^4$:

$$G(j1.57 \times 10^4) = \frac{1}{(1 - 2.465) + j4.71} = 0.203\underline{/-107.3°}$$

For $f = 3500$, $\omega = 2.2 \times 10^4$:

$$G(j2.2 \times 10^4) = \frac{1}{(1 - 4.84) + j6.6} = 0.131\underline{/-120.2°}$$

The magnitudes of the fundamental and harmonics are:

$A_1 = G(j3142)(4/\pi) = (0.768)(1.273) = 0.978$

$A_3 = G(j9425)[4/(3\pi)] = (0.353)(0.424) = 0.15$

$A_5 = G(j1.57 \times 10^4)[4/(5\pi)] = (0.203)(0.255) = 0.052$

$A_7 = G(j2.2 \times 10^4)[4/(7\pi)] = (0.182)(0.131) = 0.024$

The total distortion from the first three unwanted harmonics is:

$$D = (A_3^2 + A_5^2 + A_7^2)^{1/2}$$
$$= 0.16$$
$$\text{Percent distortion} = \frac{0.16}{0.978} = 16.4\%$$

CHECK THE SOLUTION: A numerical check is the only realistic check for this problem.

PROBLEM 18

GIVEN: The circuit of Fig. 2-30.

FIND:

   a. The Thevenin equivalent circuit to the left of the dashed line.

   b. The current and power in the load resistor for $R_L = 10\ \Omega$.

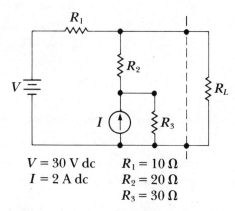

$V = 30\ \text{V dc}$ $R_1 = 10\ \Omega$
$I = 2\ \text{A dc}$ $R_2 = 20\ \Omega$
                          $R_3 = 30\ \Omega$

Fig. 2-30   Circuit configuration for Prob. 18.

c.   The value of R_L which will absorb maximum power, and determine the value of the maximum power.

SOLUTION:

a.   For Thevenin's equivalent impedance we open all current sources and short all voltage sources. The circuit becomes that of Fig. 2-31.

$$Z_{TH} = R_1\|(R_2 + R_3) = 10\|50 = 8.33\ \Omega$$

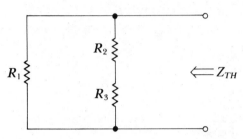

Fig. 2-31   Finding Thevenin's equivalent impedance for Prob. 18.

Using superposition, determine the voltage that exists across the output terminals with R_L removed. Do the voltage source first; open the current source branch as shown in Fig. 2-32.

$$V_{o1} = \frac{(R_2 + R_3)V}{R_1 + R_2 + R_3} = \frac{(50)(30)}{60} = 25\ \text{V dc}$$

Determine the voltage at the output terminal resulting from I alone, as shown in Fig. 2-33. Let

$$R_T = R_3\|(R_1 + R_2) = 30\|30 = 15\ \Omega$$
$$V_X = I_T R_T = (2)(15) = 30\ \text{V dc}$$
$$V_{o2} = \frac{R_1 V_X}{R_1 + R_2} = \frac{(10)(30)}{30} = 10\ \text{V dc}$$

Now the total output voltage V_{TH} due to both sources is

$$V_{TH} = V_{o1} + V_{o2} = 25 + 10 = 35\ \text{V dc}$$

The Thevenin equivalent circuit is now that shown in Fig. 2-34.

b.   The current in the load for $R_L = 10$ is

$$I_L = \frac{V_{TH}}{R_L + Z_{TH}} = \frac{35}{18.33} = 1.91\ \text{A dc}$$

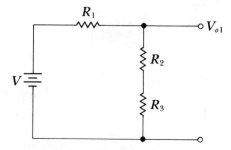

Fig. 2-32   Finding V_{o1} due to the voltage source for Prob. 18.

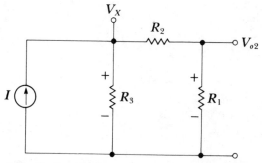

Fig. 2-33   Finding V_{o2} due to the current source for Prob. 18.

The power in the load is

$$P = I_L^2 R_L = 36.5 \text{ W}$$

c.   For maximum power transfer we want $R_L = Z_{TH}$. At $R_L = Z_{TH}$, half the voltage is across the load. Therefore the maximum power dissipated is

$$P = (V_{TH}/2)^2/R_L = (35/2)^2/8.33 = 36.76 \text{ W}$$

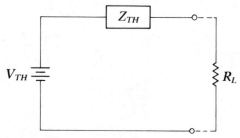

Fig. 2-34   Thevenin's equivalent circuit of Prob. 18.

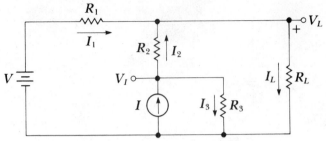

Fig. 2-35 Check circuit for Prob. 18.

CHECK THE SOLUTION: For $R_L = 10\ \Omega$, we have the circuit of Fig. 2-35 and we have determined I_L to be 1.91 A. Therefore the voltage V_L across R_L is

$$V_L = I_L R_L = 19.1 \text{ V dc}$$

Then

$$I_1 = \frac{V - V_L}{R_1} = \frac{30 - 19.1}{10} = 1.09 \text{ A dc}$$

$$I_2 = I_L - I_1 = 0.82 \text{ A dc}$$

$$V_I = I_2 R_2 + V_L = (0.82)(20) + 19.1 = 35.5 \text{ V dc}$$

$$I_3 = \frac{V_I}{R_3} = \frac{35.5}{30} = 1.18 \text{ A dc}$$

For this to check, $I = I_2 + I_3 = 2$ A dc.

$$I = 0.82 + 1.18 = 2 \text{ A dc}$$

and this checks. Let's check our maximum power transfer. Check on either side of 8.33 Ω. At $R_L = 8.6\ \Omega$,

$$I = \frac{35 \text{ V}}{8.6 + 8.33} = 2.067 \text{ A dc}$$

and the power is

$$P = I^2 R = (2.067)^2 8.6 = 36.74 \text{ W}$$

At $R_L = 8.0\ \Omega$,

$$I = \frac{35}{8.33 + 8} = 2.12 \text{ A dc}$$

and the power is

$$P = I^2 R = (2.12)^2 (8) = 36 \text{ W}$$

and the power is less for values of R either side of 8.33 Ω, thus our check is complete.

PROBLEM 19

GIVEN: It is desired to synthesize a network using an operational amplifier, resistors, and capacitors. We are given part of the circuit, as shown in Fig. 2-36. The response desired is shown in Fig. 2-37.

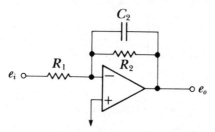

Fig. 2-36   Circuit configuration for Prob. 19.

FIND:

a.   What frequency is f_b?

b.   Where would a capacitor C_1 be located to create f_b?

c.   What are the values of R_1, C_1, R_2, and C_2 if the low-frequency input impedance is 10 kΩ?

SOLUTION:

a.   Bode analysis may be used in the analysis of this problem. Since the gain decreases at 20 dB/decade (factor of 10) or 6 dB/octave (a

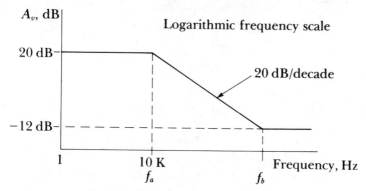

Fig. 2-37   Desired response for Prob. 19.

factor of 2), f_b can be found as follows: The first decrease of 20 dB of gain puts us at 100 kHz, while the next 6 dB puts us at 200 kHz, followed by another 6 dB which puts us at 400 kHz—a total of 32 dB [20 dB − (−12 dB)].

b. A pole exists at f_a and surely must be caused by the R_2-C_2 combination, because

$$e_o/e_{in} = -Z_2/R_1$$

where

$$Z_2 = [R_2(1/C_2s)]/(R_2 + 1/C_2s)$$

$$Z_2 = (1/C_2)/(s + 1/R_2C_2)$$

Therefore,

$$e_0/e_{in} = \frac{1}{R_1C_2}\frac{1}{s + 1/R_2C_2} \qquad \text{and} \qquad \omega_a = 1/R_2C_2$$

The zero which occurs at f_b can be synthesized similarly by locating a capacitor in shunt with R_1 as shown in Fig. 2-38.

c. The zero occurs at f_b; therefore

$$\omega_b = 1/R_1C_1$$

Also, $R_1 = 10\text{ k}\Omega$ since R_1 is the low-frequency input impedance. Knowing R_1, solve for C_1:

$$C_1 = \frac{1}{2\pi fR_1} = \frac{1}{2\pi(400 \times 10^3)(10 \times 10^3)}$$
$$= 3.98 \times 10^{-11}$$

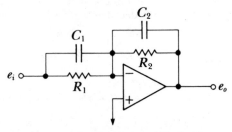

Fig. 2-38   Circuit solution to Prob. 19.

Check: At $f = 400$ kHz, X_{C1} should equal 10 kΩ, since $X_{C1} = R$.

$$X_{C1} = \frac{1}{2\pi fC_1} = \frac{1}{2\pi(400 \times 10^3)(3.98 \times 10^{-11})}$$
$$= 10\text{ k}\Omega$$

R_2 and C_2 can be synthesized now. Since the dc gain is 10 (20 dB), we can find R_2.

$$A_v = \frac{R_2}{R_1} \quad \text{at dc}$$

$$R_2 = A_v R_1 = (10)(10 \times 10^3) = 100 \text{ k}\Omega$$

From before, we know that the pole of the circuit is determined by the R_2-C_2 combination:

$$\omega_a = \frac{1}{R_2 C_2}$$

$$C_2 = \frac{1}{\omega_a R_2} = \frac{1}{2\pi(10 \times 10^3)(100 \times 10^3)}$$
$$= 1.59 \times 10^{-10}$$

Check: At $f = 10$ kHz, $X_{C2} = 100$ kΩ, since $X_{C2} = R$.

$$X_{C2} = \frac{1}{2\pi f C_2} = \frac{1}{2\pi(10 \times 10^3)(1.59 \times 15^{-10})}$$
$$= 100 \text{ k}\Omega$$

CHECK THE SOLUTION: Scan Appendix K for the Bode plot. (K1-6) satisfies this requirement. Now go to Appendix J and select a network that satisfies our transfer function:

$$A_v(s) = \frac{e_o}{e_{\text{in}}} = \frac{-Z_{T2}}{Z_{T1}} = \frac{100(\tau_b s + 1)}{(\tau_a s + 1)}$$

Select (J1-2) for Z_{T1} since Z_{T2} is already designed. Solve for τ_a and τ_b:

$$\tau_a = \frac{1}{\omega_a} = 15.92 \ \mu\text{s} \quad \text{and} \quad \tau_b = \frac{1}{\omega_b} = 0.398 \ \mu\text{s}$$

From the table for Z_{T2}:

$$A = R_2 = 100 \text{ k}\Omega, C_2 = 1.592 \times 10^{-10} \text{ F (vs. } 1.59 \times 10^{-10} \text{ F)}$$

For Z_{T1},

$$A = R_1 = 10 \text{ k}\Omega, C_1 = 3.98 \times 10^{-11} \text{ F (vs. } 3.98 \times 10^{-11})$$

PROBLEM 20

GIVEN: The three-terminal network shown in Fig. 2-39 was tested and found to have the following characteristics:

a. A dc voltage was connected to the input, and the output was loaded with 200 Ω. The output measured 5 V dc when 10 V dc was applied. When the output was short-circuited, the output current read 100 mA dc.

b.   A 6-V ac signal was applied to the input and the output was recorded for various frequencies.

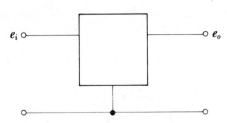

e_i o——————□——————o e_o

Fig. 2-39   Three-terminal network for Prob. 20.

Frequency	e_o	e_o/e_i
10 Hz	4.000 V rms	0.667
100 Hz	3.980 V rms	0.667
1 kHz	2.829 V rms	0.472
10 kHz	0.398 V rms	0.066
100 kHz	0.040 V rms	0.007

FIND: A circuit that meets these requirements.

SOLUTION: Part *a* gives us information concerning the dc parameters of the circuit. From this data, we can assume that the circuit is of the form shown in Fig. 2-40, with Z_1 and/or Z_2 possibly being complex impedances. At dc, we can determine the resistive components of Z_1 (R_1) and Z_2 (R_2). Let $R_2' = R_2 \| R_L$, where R_L is the 200-Ω external load resistor.

$$\frac{R_2'}{R_2' + R_1} = \frac{5\text{ V}}{10\text{ V}} = 0.5$$

Our 100-mA short-circuit measurement tells us the value of R_1:

$$\frac{10\text{ V}}{R_1} = 100\text{ mA}$$

$$R_1 = 100\ \Omega$$

Now, we can solve for R_2':

$$R_2' = 0.5R_2' + 50$$
$$= 100\ \Omega$$

Solve for R_2:

$$R_2' = \frac{R_2 R_L}{R_2 + R_L} = \frac{R_2(200)}{R_2 + 200} = 100\ \Omega$$

$$R_2 = 200\ \Omega$$

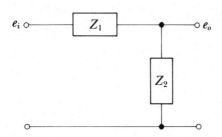

Fig. 2-40   Proposed solution circuit for Prob. 20.

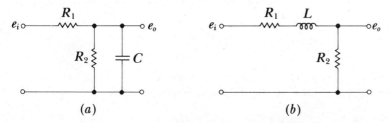

Fig. 2-41   Two possible solutions to Prob. 20.

From part b, we can see that we have a 3-dB point at 1 kHz. Also, the fact that there is 20 dB (10:1) change in gain from 10 kHz to 100 kHz tells us we have a single-pole roll-off of 20 dB/decade. Two possible solutions to the problem are shown in Fig. 2-41. Choose the circuit shown as Fig. 2-41a. From (K1-2) in Appendix K we can obtain the same circuit and the defining constants:

$$A = \frac{R_2}{R_1 + R_2} = \frac{200}{100 + 200} = 0.667 \qquad \text{(K1-2)}$$

$$T = \frac{R_1 R_2 C_1}{R_1 + R_2} = \frac{(100)(200)(C_1)}{100 + 200} = 66.7 C_1 \qquad \text{(K1-2)}$$

Also,

$$T = 1/\omega_{(3\ dB)} = 1/(2\pi)(10^3) = 1.59 \times 10^{-4}\ s$$

Solving for C, we find

$$C = T/66.7 = 2.38\ \mu F$$

CHECK THE SOLUTION: Derive the expression for $G(s) = e_a(s)/e_i(s)$ for the circuit of Fig. 2-41a:

$$G(s) = [R_2/(R_1 + R_2)] \times \{1/[R_1 R_2 Cs/(R_1 + R_2) + 1]\}$$

which checks T and A. When 10 V is applied and the output is shorted,

$$I = 10/100 = 100\ \text{mA (vs. 100 mA)}$$

Check $G(j\omega)$ at 1 kHz and 100 kHz.

$$G(j1000) = 0.472\underline{/-45°}\ \text{(vs. 0.472)}$$
$$G(j10^5) = 0.007\underline{/84.3°}\ \text{(vs. 0.002)}$$

PROBLEM 21 _____

GIVEN: It is desired to synthesize a filter with the following transfer function:

$$G(s) = \frac{(s + 10)^2(s + 5000)^2}{(s + 100)^2(s + 500)^2}$$

FIND: Sketch this waveform on a Bode plot and describe the type of filter this is. What are the magnitudes of the slope?

SOLUTION: The gain at dc is

$$G(0) = \frac{10^2(5000)^2}{100^2(500)^2} = 1 \qquad (0\ \text{dB})$$

The Bode plot is shown in Fig. 2-42 and reveals a bandpass response. We see two zeros at the first breakpoint (10 rad/s) that causes a rise of 40 dB/decade. At our next break frequency, 100 rad/s, we have two poles which cause a 40 dB/decade decrease, This causes the curve to flatten out, canceling the zeros created at 10 rad/s. At our next break frequency, 500 rad/s, we have two poles which cause a 40 dB/decade decrease; and then at 5 kHz, we have two zeros that create a 40-dB increase and cause the curve to flatten out again to zero dB.

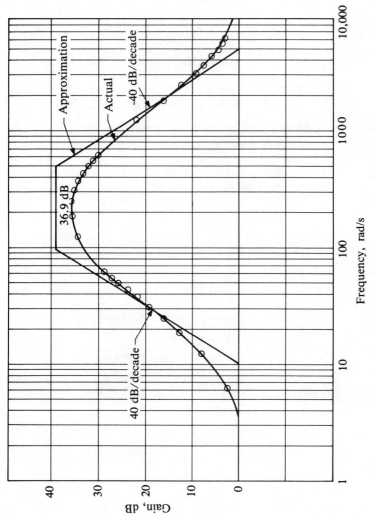

Fig. 2-42 Bode plot for Prob. 21.

119

The midband frequency can be determined by

$$f_{mid} = [(100)(500)]^{1/2} \approx 224 \text{ rad/s} \qquad (D4\text{-}3)$$

What is the gain at this frequency?

$$G(j\omega) = \frac{(j\omega + 10)^2(j\omega + 5000)^2}{(j\omega + 100)^2(j\omega + 500)^2}$$

$$|G(j224)| = \left| \frac{(j224 + 10)^2(j224 + 5000)^2}{(j224 + 100)^2(j224 + 500)^2} \right|$$

$$\approx \frac{224^2 \times 5050^2}{245^2 \times 548^2} \approx 69.7 \ (36.9 \text{ dB})$$

Many active filter books provide circuit configurations that will yield this transfer function, but let's use Appendices J or K of this book to synthesize the network.

Two stages would simplify the design. Candidates for the circuit are described by the circuit of Appendix K1-27 or the circuit described by Appendix J-13. Let's separate the filter into two identical stages as follows:

$$G_1(s) \times G_2(s) = \frac{(s + 10)(s + 5000)}{(s + 100)(s + 500)} \times \frac{(s + 10)(s + 5000)}{(s + 100)(s + 500)}$$

From Eq. (J-13), choose:

$$Z_{T2} = (s + 10)(s + 5000)$$
$$= 5 \times 10^4(0.1s + 1)(2 \times 10^{-4} + 1) \qquad (1)$$
$$Z_{T1} = (s + 100)(s + 500)$$
$$= 5 \times 10^4(0.01\,s + 1)(2 \times 10^{-3}s + 1) \qquad (2)$$

Solve for Z_{T2}: Choose network (J1-13) and solve for the component values. For Eq. (1):

$$T_1 = 2 \times 10^{-4}, T_2 = 10^{-1}, A = 5 \times 10^4$$

$$R_1 = \frac{A(T_2 - T_1)}{2T_2} = \frac{5 \times 10^4(1000 \times 10^{-4} - 2 \times 10^{-4})}{2 \times 10^{-1}}$$

$$R_1 \approx 25 \text{ k}\Omega$$

$$R_2 = \frac{AT_1}{T_2} = \frac{(5 \times 10^4)(2 \times 10^{-4})}{10^{-1}} = 100$$

$$C = \frac{2T_2^2}{A(T_2 - T_1)} = \frac{2 \times 10^{-2}}{(5 \times 10^4)(9.98 \times 10^{-2})} = 4 \ \mu\text{F}$$

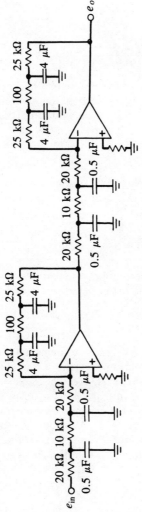

Fig. 2-43 Solution to Prob. 21.

121

Solve for Z_{T1}: $T_1 = 2 \times 10^{-3}$, $T_2 = 10^{-2}$, $A = 5 \times 10^4$

$$R_1 = \frac{(5 \times 10^4)(10^{-2} - 2 \times 10^{-3})}{2 \times 10^{-2}} = 20 \text{ k}\Omega$$

$$R_2 = \frac{AT_1}{T_2} = \frac{5 \times 10^4(2 \times 10^{-3})}{10^{-2}} = 10 \text{ k}\Omega$$

$$C = \frac{2 \times 10^{-4}}{5 \times 10^4(8 \times 10^{-3})} = 0.5 \text{ } \mu\text{F}$$

The circuit configuration combines the two stages and is shown in Fig. 2-43. Use the closest standard values for all components. In actual practice, we pick the closest capacitor value and recalculate the resistors since they are available in 2% steps for metal films. In Eq. (J-3) we had a ($-$) sign for this gain, but two stages produce an overall transfer function that is positive.

CHECK THE SOLUTION: Check the calculations for the components and then derive the expression for the transfer function of (J1-13), i_o/e_{in}.

PROBLEM 22 _____

GIVEN: It is desired to synthesize an active network that will give either of the input/output relationships as shown in Fig. 2-44a and b.

FIND: The network that will accomplish this function and give the element values.

SOLUTION: From Fig. 2-44a, we can determine that the circuit to be synthesized is an integrator; however, the output of Fig. 2-44b integrates to 7.5 V, then ceases to increase. This can be implemented with a limiter, and a possible solution to this problem is shown in Fig. 2-45. A_1 is a unity gain inverter while A_2 performs the integrating function. The zener voltage is selected to be 7.5 V to limit the output signal at this level. Pick R_3 to be 1 kΩ and solve for C as follows:
    From Fig. 2-45, the current through R_3 is

$$i_{R3} = \frac{-e_{\text{in}}}{R_3} = \frac{-1 \text{ V}}{1 \text{ k}\Omega} = -1 \text{ mA}$$

since $e_a = -e_{\text{in}}$. This current must flow through C. We want to rise 2 V in 1 ms, and the following relationship allows us to select C. $i = C(dv/dt)$.

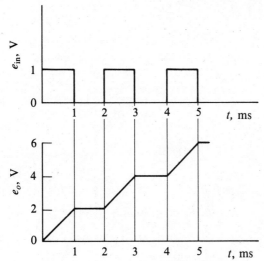

Fig. 2-44a Input/output relationship of Prob. 22.

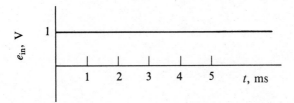

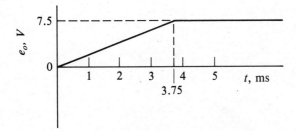

Fig. 2-44b Input/output relationship of Prob. 22.

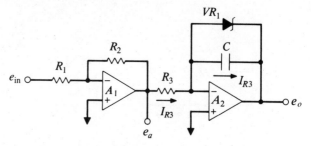

Fig. 2-45   Possible solution to Prob. 22.

$$C = i(\Delta t/\Delta V) = (10^{-3})(10^{-3})/2 = 0.5 \ \mu F$$

For A_1, choose $R_1 = R_2$ for a unity gain inverter.

CHECK THE SOLUTION: The voltage across a capacitor can be related to its current by

$$e = \frac{1}{C} \int i \ dt = \frac{1}{0.5 \times 10^{-6}} \int_0^{10^{-3}} 10^{-3} \ dt = \frac{(10^{-3})(10^{-3})}{0.5 \times 10^{-6}} = 2 \ V$$

and our check is complete.

3

SERVO CONTROL
AND FEEDBACK PROBLEMS

Most of the analysis performed in this chapter uses the Routh criterion and Bode analysis to determine system stability. Other techniques, such as the Nyquist, root-locus, and Nichols' chart are presented and any approach may be specifically requested as the solution approach on the examination.

When system performance is evaluated, again Bode analysis is normally exercised, primarily because of the familiarity of most engineers with this technique. Since almost all computer-analysis printouts are Bode, it is not an inconvenient technique with which to familiarize oneself. Bode is somewhat limited in certain time-domain analyses and other techniques may prove more efficient for some problems.

Most stability analyses are based upon the fact that if there exists a gain of 0 dB or greater at zero phase shift (or 360°), then a potentially unstable condition exists because this is the criterion for oscillation. Since all feedback systems possess an inherent 180° phase shift because of the negative feedback contribution (error amplifier), only 180° of phase shift remain for other contributions. Normally, a system is determined to be stable if it has a phase margin of 45° or more when the gain is 0 dB, or it has a gain margin of 6 dB or more when the phase is 360°.

A very helpful publication is available that is reasonably priced and contains many excellent problem solutions in the area of feedback and control systems. The publication is entitled "Feedback and Control Systems" by DiStefano, Stubberud, and Williams and is part of the Schaum Outline Series published by McGraw-Hill.

Some of the problems are more complex than those that might be given on the examination, for example, synthesis of a compensating circuit. These have been included to broaden the scope of the material.

Appendices D, E, and G to K are applicable to solving the types of problems given in this chapter. Equations are represented in the text as follows: (G-1) means the equation is taken from Appendix G and is equation 1.

PROBLEM 1

GIVEN: The following loop-gain function:

$$G(s)H(s) = \frac{10(1+0.12s)}{s(1+0.1s)(1+0.03s)(1+0.04s)}$$

FIND: Check the stability using the Routh criterion.

SOLUTION: Set:

$$1 + G(s)H(s) = 0$$

$$10(1+0.12s) + s(1+0.1s)(1+0.03s)(1+0.04s) = 0$$

Multiply terms to get the fourth-order characteristic equation:

$$10 + 1.2s + (s + 0.1s^2)(1 + 0.07s + 0.0012s^2) = 0$$

$$1.2 \times 10^{-4}s^4 + 0.0082s^3 + 0.17s^2 + 2.2s + 10 = 0$$

Set up the Routh array:

s^4	0.00012	0.17	10
s^3	0.0082	2.2	0
s^2	$\dfrac{0.00113}{0.0082}$	10	0
s^1	$\dfrac{0.2212}{0.1378}$	0	0
s^0	10	0	0

Since we have no changes of sign in the first column the system is stable.

CHECK THE SOLUTION: Let's check our loop-gain function to see what the phase is at 0 dB.

$$G(s)H(s) = \frac{10(1 + 0.12s)}{s(1 + 0.1s)(1 + 0.03s)(1 + 0.04s)}$$

$$GH(j\omega) = \frac{10(1 + 0.12j\omega)}{j\omega(1 + 0.1j\omega)(1 + 0.03j\omega)(1 + 0.04j\omega)}$$

$$= \frac{10^4(8.33 + j\omega)}{j\omega(10 + j\omega)(33 + j\omega)(25 + j\omega)}$$

Let's plot this function on semilog paper as shown in Fig. 3-1 and determine from this plot at what frequency we have 0 dB. We can then substitute this value of ω into the $GH(j\omega)$ equation to get our phase. As a starting point, we need a gain point far away from a break, for accuracy purposes. Pick $j\omega = 0.1$ rad/s and determine $GH(j\omega)$:

$$GH(j0.1) = \frac{10^4(8.33 + j0.1)}{j0.1(10 + j0.1)(33 + j0.1)(25 + j0.1)} \approx -j100$$

$$|GH(j0.1)| = 40 \text{ dB}$$

From Fig. 3-1 we see that 0 dB occurs at approximately 10 rad/s. Find $GH(j\omega)$ at this frequency.

$$GH(j10) = \frac{10^4(8.33 + j10)}{j10(10 + j10)(33 + j10)(25 + j10)}$$

$$= \frac{13 \times 10^4 \underline{/50.2°}}{(10\underline{/90°}(14.1\underline{/45°})(34.5\underline{/16.9°})(26.9\underline{/21.8°})}$$

$$= \frac{13 \times 10^4 \underline{/50.2°}}{13.09 \times 10^4 \underline{/173.7°}} = 0.993\underline{/-123.5°}$$

This says we have about 56.5° (180°–123.5°) of phase margin at a gain of 0.993(−0.06 dB). Since 45° nominal is considered acceptable for stability, our check is complete.

PROBLEM 2

GIVEN: The servo system shown in Fig. 3-2.

FIND: Determine what range K must be within for the system to be stable.

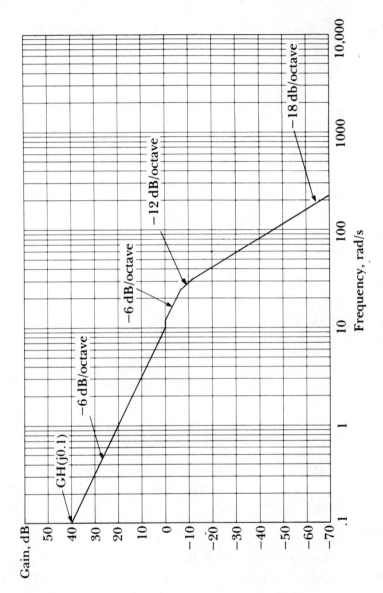

Fig. 3-1 Bode plot used in solution check of Prob. 1.

SOLUTION: Let's solve using the Routh criterion:

$$G(s) = G_1(s)G_2(s) = \frac{K}{(s+2)(s^2+4s+20)} = \frac{K}{D}$$

Set $1 + G(s)H(s) = 0$:

$$1 + \left[\frac{K}{(s+2)(s^2+4s+20)}\right]\frac{10}{s} = 0$$

$$s(s+2)(s^2+4s+20) + 10K = 0$$

$$s^4 + 6s^3 + 28s^2 + 40s + 10K = 0$$

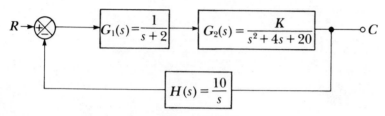

Fig. 3-2   Servo system block diagram of Prob. 2.

Set up the Routh array:

s^4	1	28	$10K$
s^3	6	40	0
s^2	$\dfrac{(6)(28)-(1)(40)}{6} = 21.33$	$\dfrac{60K-1(0)}{6} = 10K$	0
s^1	$\dfrac{(21.33)40-60K}{21.33} = 40-2.813K$	0	
s^0	$10K$		

Solve for the range of K:

$$K = 0 \qquad \text{(for the } s^0 \text{ term)}$$

$$40 - 2.813K = 0 \qquad \text{(for the } s^1 \text{ term)}$$

$$2.813K = 40$$

$$K \leqslant 14.22$$

Therefore K must lie between 0 and 14.22.

CHECK THE SOLUTION: Insert $K = 10^{-3}$ (close to zero) into the Routh array; the s^0 term becomes 0.01. This checks. Also select K a little below 14.22, say 14.21; the s^1 term becomes 0.034, and this checks.

PROBLEM 3

GIVEN: The block diagram of a servo feedback system as shown in Fig. 3-3. The transfer function for $G_2(s)$ is not known, but it has been empirically determined to be as shown in Fig. 3-4. (*Note*: Dashed lines are not given in the problem. These are drawn in during the solution to find the breakpoints.)

FIND: $C(s)/R(s)$

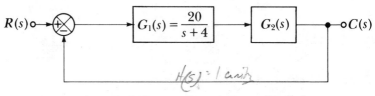

Fig. 3-3 Block diagram of servo system of Prob. 3.

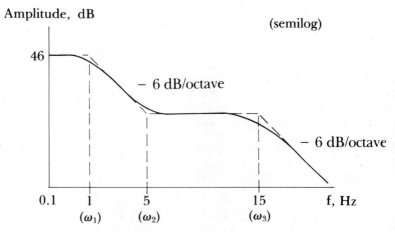

Fig. 3-4 Bode plot of $G_2(s)$ of Prob. 3.

SOLUTION: One can synthesize the transfer function by realizing that single-order poles occur at 1 Hz ($\omega_1 = 6.28$ rad/s) and at 15 Hz ($\omega_3 = 94.25$ rad/s), with a zero occurring at 5 Hz ($\omega_2 = 31.42$ rad/s). The dc gain is 200 (46 dB). The transfer function for $G_2(s)$ is

$$G_2(s) = \frac{K(s + \omega_2)}{(s + \omega_1)(s + \omega_3)}$$

$$= \frac{K(s + 31.42)}{(s + 6.28)(s + 94.25)}$$

We can find K because we know the dc gain is 200:

$$G_2(s)\Big|_{s=0} = 200 = \frac{K(31.42)}{(6.28)(94.25)}$$

$$K = 3768$$

and the total transfer function becomes:

$$G_2(s) = \frac{3768(s + 31.42)}{(s + 6.28)(s + 94.25)}$$

The total forward gain $G(s)$ can be found as follows:

$$G(s) = G_1(s)G_2(s) = \frac{7.536 \times 10^4(s + 31.42)}{(s + 4)(s + 6.28)(s + 94.25)} = \frac{N}{D}$$

and the total closed-loop transfer function can be found with the expression:

$$\frac{C(s)}{R(s)} = \frac{G(s)}{1 + G(s)H(s)} = \frac{N/D}{1 + N/D} = \frac{N}{N + D} \qquad H(s) = 1$$

$$= \frac{7.536 \times 10^4(s + 31.42)}{7.536 \times 10^4(s + 31.42) + (s + 4)(s + 6.28)(s + 94.25)}$$

CHECK THE SOLUTION: Check $C(s)/R(s)$ at $s = 0$. $C(s)/R(s) = 0.999$, which it should be for unity feedback. Check $G_1(s)$ and $G_2(s)$ at $s = 0$. $G_1(s) = 5$ and $G_2(s) = 200$. Therefore, $G(s) = 1000$, and

$$\frac{C(s)}{R(s)} = \frac{1000}{1 + 1000} = 0.999$$

Another check is to find the transfer function in Appendix K as (K1-15). From this:

$$\tau_1 = \frac{1}{\omega_1} = 0.0318 \text{ s} \qquad \tau_3 = \frac{1}{\omega_3} = 0.1592 \text{ s}$$

and

$$\tau_4 = \frac{1}{\omega_4} = 0.0106 \text{ s}$$

For these parameters, $G_2(s)$ checks:

$$G_2(s) = \frac{3768(s + 31.42)}{(s + 6.28)(s + 94.25)}$$

PROBLEM 4

GIVEN: An angular-position servo system uses potentiometer feedback and has the following gain/transfer functions:

Amplifier:

$$G_1(s) = \frac{100}{(s+5)} \frac{\text{V}}{\text{V}}$$

Motor mechanical
transfer function:

$$G_2(s) = \frac{1}{40s} \frac{\text{rad}}{\text{A}}$$

Motor electrical
transfer function:

$$G_3(s) = \frac{1}{s+4} \frac{\text{A}}{\text{V}}$$

Potentiometer gain
constant:

$$G_4(s) = K \frac{\text{V}}{\text{rad}}$$

FIND:

    a.   Draw a block diagram of this system.

    b.   Determine the open-loop gain function.

    c.   Determine the closed-loop transfer function.

    d.   If $K = 100$, is the system stable?

SOLUTION:

    a.   Our block diagram for this system is shown in Fig. 3-5.

    b.   The open-loop transfer function is

$$G(s) = G_1(s)G_2(s)G_3(s) = \frac{100}{40s(s+5)(s+4)} = \frac{2.5}{s(s+5)(s+4)}$$

    c.   The closed-loop transfer function requirements suggest that we put the system in canonical form as follows:

$$C/R = \frac{G(s)}{1 + G(s)H(s)} = \frac{2.5}{(s)(s+5)(s+4) + 2.5K}$$

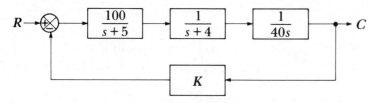

Fig. 3-5   Angular position servo system of Prob. 4.

d. The characteristic equation for this system is

$$(s)(s + 5)(s + 4) + 2.5K = 0$$

$$s^3 + 9s^2 + 20s + 2.5K = 0$$

Set up a Routh array to find the gain range of K for a stable system.

s^3	1	20	0
s^2	9	2.5K	0
s^1	$\dfrac{180 - 2.5K}{9}$	0	
s^0	2.5K	0	

Setting the s^1 and s^0 terms in the first column ≥ 0, we can solve for the range of K's that make the system stable.

For $180 - 2.5K \geq 0$: $K \leq 72$
For $2.5K \geq 0$: $K \geq 0$

so the range of K for a stable system is $0 \leq K \leq 72$ V/rad. The answer for $K = 100$ is that the system is not stable.

CHECK THE SOLUTION: Let $K = 0.01$, which is close to 0. The s^0 term becomes 0.025, which is slightly positive. For $K = 71$, the s^1 entry becomes 0.278, which is slightly positive.

PROBLEM 5

GIVEN: The analog simulation of a servo system is shown in Fig. 3-6.

FIND:

a.   The Laplace expression that represents this system.

b.   The block diagram showing the transfer function.

c.   The analog computer representation of the following expression in block form:

$$e_1(t) = 100 \, de_2(t)/dt + e_2(t)$$

Assume initial conditions are zero.

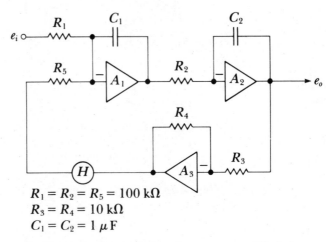

$R_1 = R_2 = R_5 = 100 \text{ k}\Omega$
$R_3 = R_4 = 10 \text{ k}\Omega$
$C_1 = C_2 = 1 \, \mu\text{F}$

Fig. 3-6   Analog simulation of servo system of Prob. 5.

SOLUTION:

b.   Do part b first since it will give us part a. Refer to Fig. 3-7 and write the expression for the gain of the first amplifier A_1:

$$i_i = \frac{e_i}{R_1} = -i_o$$

$$e_o = i_o(1/C_1s) = -e_i/R_1C_1s$$

$$\frac{e_o}{e_i} = -1/R_1C_1s = -10/s$$

For the second amplifier A_2, the derivation is similar:

$$\frac{e_o}{e_i} = -1/R_2C_2s = -10/s$$

The block diagram becomes that of Fig. 3-8 if we realize that A_3 is only an inverting amplifier with unity gain.

a.   The form and gain of the system is

$$G(s) = (-10/s)(-10/s) = 100/s^2$$

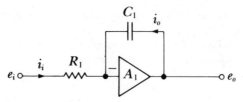

Fig. 3-7   First stage of Prob. 5.

The feedback term $H(s)$ is

$$H(s) = -H$$

The overall transfer function can be put into canonical form.

$$\frac{C}{R} = \frac{G(s)}{1 - G(s)H(s)} = \frac{100/s^2}{1 + 100H/s^2} = \frac{100}{s^2 + 100H}$$

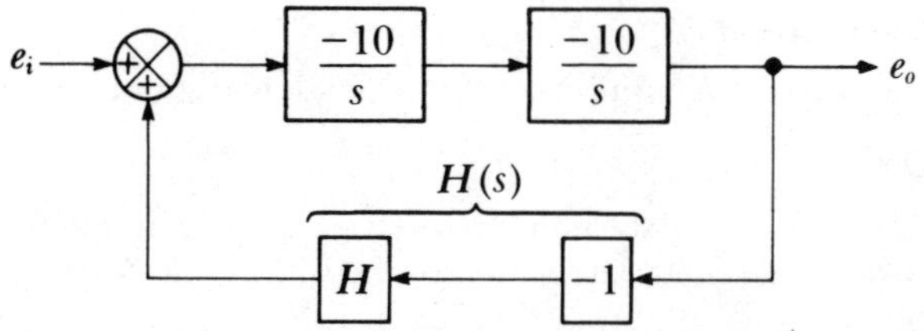

Fig. 3-8   Block diagram representation of Prob. 5.

c.   Let us represent the differential equation in Laplace notation as follows:

$$e_1(s) = 100se_2(s) + e_2(s) = (100s + 1)e_2(s)$$

$$\frac{e_2(s)}{e_1(s)} = \frac{1}{100s + 1} = \frac{G(s)}{1 + G(s)H(s)}$$

$$= \frac{1/(100s)}{1 + 1/(100s)}$$

and $G(s) = 1/(100s)$; $H(s) = 1$. The block diagram is shown in Fig. 3-9.

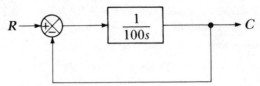

Fig. 3-9   Analog block representation of Prob. 5, part *c*.

CHECK THE SOLUTION: This problem does not lend itself well to alternate solution techniques, and the best check is to check the problem for numerical accuracy and to rederive expressions.

PROBLEM 6

GIVEN: The loop-gain function $GH(s) = K/s(s + \omega_1)(s + \omega_2)$ is represented by the straight-line approximation Bode plot shown in Fig. 3-10. The solid line represents the gain and the dashed line represents phase.

FIND:

    a. ω_1 and ω_2

    b.   *K*

    c.   Phase margin ϕ_m

    d.   The value of *K* adjustment for a phase margin of 45°

SOLUTION:

    a.   The first pole determined by $1/s$ is at zero frequency. The other two breakpoints can be determined from Fig. 3-10 to be

$$\omega_1 = 10 \text{ rad/s} \qquad T_1 = 1/\omega_1 = 0.1 \text{ s}$$
$$\omega_2 = 100 \text{ rad/s} \qquad T_2 = 1/\omega_2 = 0.01 \text{ s}$$

    b.   The transfer function is now:

$$GH(s) = \frac{K'}{s(0.1s + 1)(0.01s + 1)} = \frac{K}{s(s + 10)(s + 100)}$$

$$GH(j\omega) = \frac{K}{j\omega(j\omega + 10)(j\omega + 100)}$$

Since we have a straight-line approximation, our maximum gain error occurs at the breakpoints. To solve for *K* with minimum error, let's move as far from a break as possible. Choose $\omega = 0.1$ rad/s.

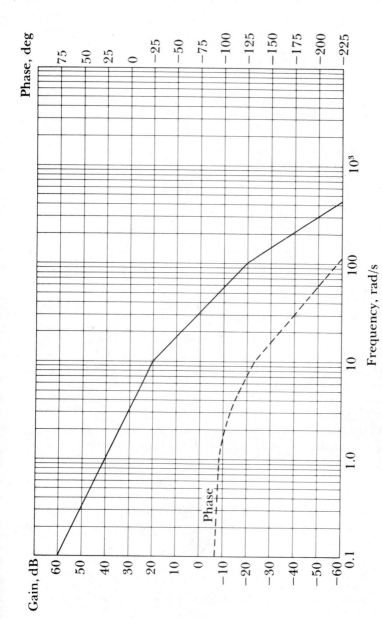

Fig. 3-10 Straight-line Bode plot approximation for Prob. 6.

137

$$|GH(j\omega)| = 60 \text{ dB} = 1000 = \left|\frac{K}{(0.1j)(0.1j + 10)(0.1j + 100)}\right|$$

$$\left|\frac{K}{(0.1\underline{/90°})(10\underline{/0.6°})(100\underline{/0.06°})}\right| = 1000$$

$$\frac{K}{100} = 1000$$

$$K = 10^5$$

c. From the figure, we can determine that the phase is $-175°$ at the point where the gain is 0 dB. Our phase margin is

$$\phi_m = 180° - 175° = 5°$$

This would be very marginal for stability, since we would prefer 45°.

d. If we want a phase margin of 45°, the phase plot has a value of:

$$\phi = -180° + \phi_m = -180° + 45° = -135°$$

At this value of ϕ on Fig. 3-10, our gain is 20 dB at a frequency of 10 rad/s. We would like our gain to be 0 dB at this magnitude of phase and frequency. Therefore, we would reduce our gain 20 dB (a factor of 10). K now becomes 10^4.

CHECK THE SOLUTION: Check the gain at 400 rad/s. $GH(j400) = 1.5 \times 10^{-3}\underline{/-254.6°}$. $20 \log(1.5 \times 10^{-3}) = -56.5$ dB (vs. -56 dB). Now calculate $GH(j10) = 0.704\underline{/-140.7°}$, which is close to the 0.707 expected.

PROBLEM 7

GIVEN: Two transfer functions, as shown in Fig. 3-11, which are straight-line approximations of the gain characteristics with frequency.

FIND:

a. The transfer function for G_1 and G_2.

b. The transfer function for their product.

c. If the two gain functions are in the forward loop of a unity feedback system, is the system stable?

SOLUTION:

a. In the solution, we will assume that G_1 has a pole at zero, since the plot does not show anything below 0.1 rad/s. Also, for G_1 we see 20 dB/decade roll-off up to a break at 50 rad/s and 40 dB/decade

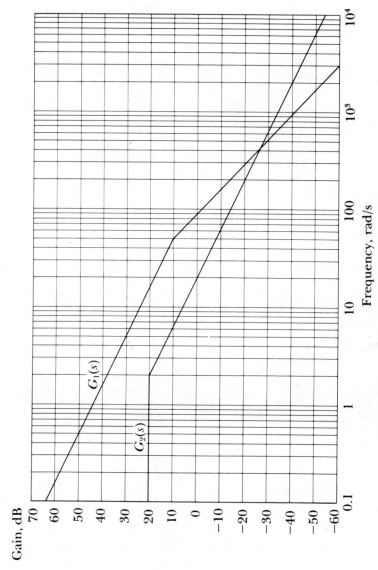

Gain, dB

Frequency, rad/s

Fig. 3-11 $G_1(s)$ and $G_2(s)$ Bode plot representation of Prob. 7.

past 50 rad/s. Therefore, G_1 is of the form

$$G_1(s) = \frac{A_1}{s(\tau_1 s + 1)}$$

where

$$\tau_1 = \frac{1}{\omega_1} = \frac{1}{50} = 0.02 \text{ s}$$

G_1 has a magnitude of 50 dB at 0.5 rad/s; therefore, we can solve for A_1. (*Note*: We should pick a gain point remote from any breakpoint for accuracy, since we use a straight-line approximation.)

$$|G_1(j0.5| = \left| \frac{A_1}{j0.5[(0.02)(0.5j) + 1]} \right| = 50 \text{ dB} = 316.2$$

$$\left| \frac{A_1}{j0.5} \right| \approx 316.2$$

$$A_1 = 158.1$$

and our transfer function for G_1 is

$$G_1(s) = \frac{158.1}{s(0.02s + 1)}$$

Now let's solve for G_2, which is a simple first-order roll-off since it changes at a rate of -20 dB/decade beyond 2 rad/s. Our transfer function is of the form

$$G_2(s) = \frac{A_2}{\tau_2 s + 1}$$

$$\tau_2 = 1/\omega_2 = 0.5$$

and A_2 is nothing more than the gain at dc or when $s = 0$. This value is 20 dB or $A_2 = 10$.

Therefore, our transfer function for A_2 is

$$G_2(s) = \frac{10}{0.5s + 1}$$

b.   The total gain $G(s)$ is equal to $G_1(s) \times G_2(s)$:

$$G(s) = \frac{1581}{s(0.02s + 1)(0.5s + 1)}$$

c.   One might suspect that the system is unstable since we have a slope of -60 dB/decade past $\omega = 50$ rad/s. Let's make a Bode plot as shown in Fig. 3-12. We obtain this by adding the dB gains at each

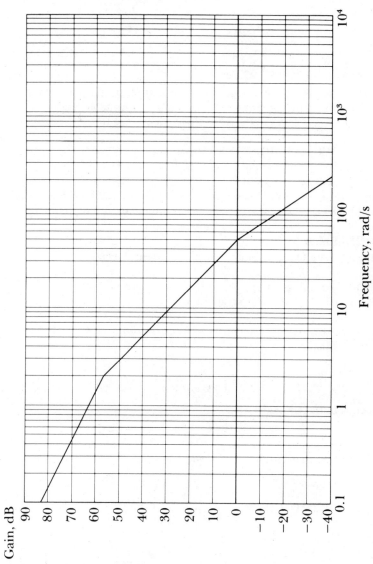

Gain, dB

Frequency, rad/s

Fig. 3-12 $G_1(s)$ and $G_2(s)$ total transfer function.

141

frequency. We can see that we cross $0\,dB$ at a frequency of approximately $50\,rad/s$. Calculate the gain and phase at this frequency.

$$G(j50) = \frac{1581}{j50(j1+1)(j25+1)} \approx \frac{1581}{(50\underline{/90°})(1.41\underline{/45°})(25\underline{/87.7°})}$$

$$= 0.893\underline{/-222.7}$$

which says we have a negative phase margin of $42.7°$ at a gain of $0.893(-0.98\,dB)$, and the system is definitely unstable, inasmuch as we are interested in $GH(j\omega) = G(j\omega)$, since $H = 1$.

CHECK THE SOLUTION: Check $G_1(s)$ at $\omega = 10^3\,rad/s$. $|G_1| = -42$ dB, which is 0.0079 from the curves. $|G_1(j10^3)| = 7.905 \times 10^{-3}$ calculated. For $G_2(s)$, check at $2000\,rad/s$. $|G_2(j2000)| = 0.01$, which corresponds to -40 dB. One other check is to use Routh's criterion. For this problem, s^1 becomes -204, demonstrating an unstable condition.

PROBLEM 8
GIVEN: The block diagram of a servo system shown in Fig. 3-13.

FIND: The values of A and a from the expression $A/s(s+a)$ for a 15% overshoot and a 10-s settling time.

SOLUTION: We must write the transfer function for the closed-loop system, which is

$$F(s) = \frac{C(s)}{R(s)} = \frac{G(s)}{1 + G(s)H(s)} = \frac{A}{s^2 + as + A}$$

which is in the classic form of

$$F(s) = \frac{\omega_n^2}{s^2 + 2\xi\omega_n s + \omega_n^2}$$

Let's assume that settling time means less than 5%, which it normally

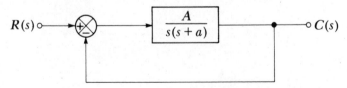

Fig. 3-13   Block diagram of servo system of Prob. 8.

implies. The equation for a 5% settling time can be found in many texts as

$$t_s = \frac{3}{\xi \omega_n}$$

and the percent overshoot can be found to be:

$$PO = 100 \exp[(-\xi\pi)/(1 - \xi^2)^{1/2}]$$

Solving for the setting time criteria, we find

$$\xi \omega_n = 3/10 = .3$$

And from the curve in Fig. 3-14 we determine that the ξ which satisfies our requirements is approximately 0.53 for 15% of overshoot. Therefore,

$$\omega_n = 0.3/\xi = 0.57$$
$$\omega_n^2 = 0.32$$

and our closed-loop expression becomes:

$$F(s) = \frac{0.32}{s^2 + 0.6s + 0.32}$$

$$A = 0.32$$
$$a = 0.6$$

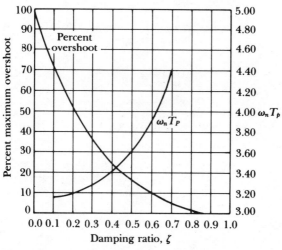

Fig. 3-14 Peak overshoot and peak time versus damping ratio for a second-order system. This or similar graphs can be found in many texts.

CHECK THE SOLUTION: For our overshoot,

$$PO = 100 \exp{[-0.53\pi/(0.72)^{1/2}]} = 14.12\%$$

which is fairly close. We can check our settling time as follows:

$$t_s = \frac{3}{\xi\omega_n} = \frac{3}{(0.53)(0.57)}$$
$$= 9.93 \text{ s}$$

and our check is complete.

PROBLEM 9

GIVEN: The block representation of a servo system with a loop delay represented by e^{-sT}, as shown in Fig. 3-15.

FIND: The maximum value of T for a stable system.

SOLUTION: The loop gain should be found first.

$$GH(s) = G(s)H(s) = G_1(s)G_2(s)G_3(s)e^{-sT}$$

Neglecting the phase shift, we obtain $GH'(s)$:

$$GH'(s) = \frac{10}{s+1} \cdot \frac{40}{s+0.1} \cdot \frac{s}{s+5} = \frac{400s}{(s+0.1)(s+1)(s+5)}$$

$$GH'(j\omega) = \frac{400j\omega}{(0.1+j\omega)(1+j\omega)(5+j\omega)}$$

For a starting point, let's find $G(j\omega)$ far from a breakpoint for accuracy purposes and plot the function. Pick $\omega = 0.01$ rad/s and solve for $GH(j\omega)$:

$$GH(j0.01) = \frac{4j}{(0.1)(1)(5)} = 8j$$

$$\left| GH(j0.01) \right| = 20 \log_{10} 8 = 18.06 \text{ dB}$$

Now we can use straight-line approximations as shown in Fig. 3-16.

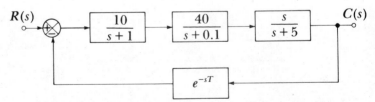

Fig. 3-15   Block representation of servo system for Prob. 9.

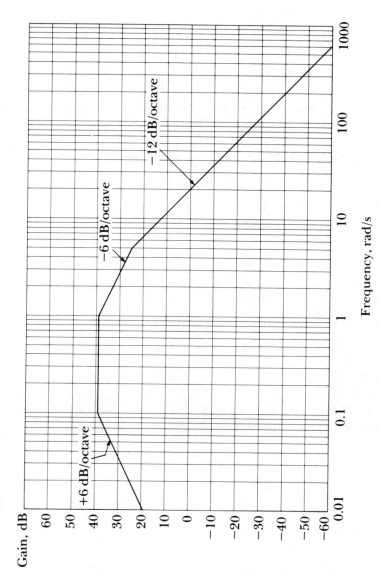

Fig. 3-16 Bode plot straight-line approximation for $GH(s)$ of Prob. 9.

145

It appears as though we cross 0 dB at 20 rad/s. What is our phase at this crossover point?

$$GH(j20) = \frac{400(j20)}{(0.1 + j20)(1 + j20)(5 + j20)} \approx \frac{400}{(20\underline{/87.1°})(20.6\underline{/76°})}$$

$$GH(j20) = 0.97\underline{/-163.1°}$$

The gain checks since it is close to 1. Our phase margin is only 16.9° (180°–163.1°) and is already potentially unstable. It will surely be unstable if we add another 16.9°. Therefore, at $\omega = 20$ rad/s,

$$\omega T = 16.9° \times (2\pi \text{ rad}/360°)$$
$$\omega T = 0.294 \text{ radian}$$
$$T = 0.294/20 = 14.75 \text{ ms}$$

CHECK THE SOLUTION. The problem as worked is basically self-checking. However, let's make a gain check at $\omega = 0.5$ rad/s.

$$GH(j0.5) = \frac{400(j0.5)}{(0.1 + j0.5)(1 + j0.5)(5 + j0.5)}$$

$$\approx \frac{200\underline{/90°}}{(0.51\underline{/78.7°})(1.12\underline{/26.6°})(5\underline{/5.7°})} = 70\underline{/-21°}$$

$$|GH(j0.5)| = 20 \log_{10} 70 = 37 \text{ dB}$$

This checks fairly closely with the Bode plot.

PROBLEM 10 _____

GIVEN: The block diagram of a servo system is shown in Fig. 3-17. We want the system closed-loop transfer function $G(s) = C(s)/R(s)$ to possess the following characteristics: $\xi = 0.707$, $\omega_n = 15$.

FIND:

    a. The transfer function of a compensating network to accomplish this using cancellation compensation.

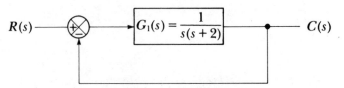

Fig. 3-17   Servo system block diagram of Prob. 10.

b.   The network required to provide this function.

SOLUTION: Cancellation compensation is accomplished by removing a pole or zero and replacing it with another pole or zero. For example, if we have a function of the form $G(s) = 1/(s + a)$ and we desire $G'(s) = 1/(s + b)$ we can multiply our original function by $G_C(s) = (s + a)/(s + b)$, the compensating network. Thus, we can get our desired function $G'(s)$ as follows:

$$G'(s) = G(s)G_C(s) = [1/(s + a)](s + a)/(s + b) = 1/(s + b)$$

a.   The closed-loop transfer function can be found from:

$$G(s) = \frac{C(s)}{R(s)} = \frac{G_1(s)}{1 + G_1(s)H(s)}$$

Since $H(s) = 1$ and $G_1(s) = 1/s(s + 2)$:

$$G(s) = \frac{1}{s^2 + 2s + 1}$$

This is of the general form of a second-order system described by:

$$G(s) = \frac{\omega_n^2}{s^2 + 2\xi\omega_n^s + \omega_n^2} \qquad (G2\text{-}1)$$

For the circuit of Fig. 3-16, we have

$$\omega_n = 1$$
$$\xi = 1$$

For the given requirements of $\xi = 0.707$ and $\omega_n = 15$ we have $\omega_n^2 = 225$ and $2\xi\omega_n = 21.21$, and our desired closed-loop function becomes:

$$G'(s) = \frac{225}{s^2 + 21.21s + 225}$$

Now we can take this expression back to open-loop form and obtain $G_1'(s)$ from

$$G'(s) = \frac{G_1'(s)}{1 + G_1'(s)H(s)} = \frac{N_1'(s)}{H(s)N_1'(s) + D_1'(s)}$$

where $G_1'(s) = \frac{N_1'(s)}{D_1'(s)}$

$$= \frac{225}{s^2 + 21.21s} = \frac{225}{s(s + 21.21)}$$

If we use cancellation techniques, we should cancel out our original pole with a zero and insert a new pole as described by $G_1'(s)$. Since our gain is 225, our compensation network now becomes

$$G_C(s) = \frac{225(s+2)}{s+21.21}$$

Our new forward-loop transfer function is now

$$G_1'(s) = G_1(s)G_C(s)$$

$$= \left[\frac{1}{s(s+2)}\right]\left[\frac{225(s+2)}{s+21.21}\right] = \frac{225}{s(s+21.21)}$$

b. We can synthesize a transfer function or we can use the networks of Table J1 for an active-circuit solution or the circuits of Appendix K for a passive solution. Let's use an active solution because we get good isolation (low output impedance). Let's get our compensating network in the form described by Table J1. The gain will be negative because of the inverting amplifier.

$$G_C(s) = \frac{-225(s+2)}{s+21.21} = \frac{-21.21(0.5s+1)}{0.047s+1}$$

These requirements can be satisfied by the circuits of Tables J1-5a, J1-5b, and J1-5c. Choose J1-5b. The transfer-impedance function is

$$G_C(s) = \frac{A(1+sT)}{(1+s\theta T)}$$

where $A = 21.22$

$$T = 0.5$$

$$\theta T = 0.047$$

$$\theta = 0.094$$

From Table J1-5b,

$$R_1' = A/2 = 21.22/2 = 10.61$$

$$R_2' = A\theta/[4(1-\theta)] = 0.55$$

$$C = 4T(1-\theta)/A = 0.0854$$

Choose $C = 100\ \mu F$ and scale the other values accordingly.

$$\alpha = 0.0854/10^{-4} = 854$$

$$R_1 = \alpha R_1' = 9.06\ k\Omega$$

$$R_2 = \alpha R_2' = 470\ \Omega$$

We would choose the nearest standard value for R_1 of $9.09\ k\Omega$ from Appendix H.

The dc gain for the compensator is

$$G_C(s) = -21.22$$
$$\scriptstyle s\to 0$$

Therefore, since C is an open circuit at dc, our dc gain function is given in Appendix C2 as

$$|A| = \frac{Z_f}{Z_i} \qquad (C2\text{-}1)$$

$$\frac{2R_1}{R_3} = \frac{18.12 \times 10^3}{R_3}$$

$$R_3 = 854 \ \Omega$$

Choose the nearest standard value of 845 Ω from Table H1-1. Note that we have introduced 180° of phase reversal when we use an inverting amplifier. We must therefore change the sign of the summer at the input error circuit. The circuit implementation is shown in Fig. 3-18 and the system block diagram in Fig. 3-19.

CHECK THE SOLUTION:

a.   Check the new closed-loop gain function:

$$G_1'(s) = G_C(s)G_1(s) = \left[\frac{-21.22(0.5s + 1)}{0.047s + 1} \right]\left[\frac{1}{2s(0.5s + 1)} \right]$$

$$= \frac{-21.22}{0.094s^2 + 2s} = \frac{-225}{s^2 + 21.28s}$$

$$G'(s) = \frac{-G_1'(s)}{1 - G_1'(s)H(s)} = \frac{225}{s^2 + 21.28s + 225}$$

and this checks.

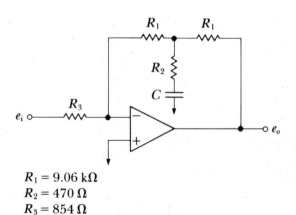

$$R_1 = 9.06 \ \text{k}\Omega$$
$$R_2 = 470 \ \Omega$$
$$R_3 = 854 \ \Omega$$
$$C = 100 \ \mu\text{F}$$

Fig. 3-18   Cancellation compensation circuit of Prob. 10.

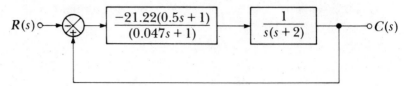

$$R(s) \circ\!\!-\!\!\bigotimes\!\!-\!\!\boxed{\dfrac{-21.22(0.5s+1)}{(0.047s+1)}}\!\!-\!\!\boxed{\dfrac{1}{s(s+2)}}\!\!-\!\!\bullet\!\!-\!\!\circ C(s)$$

Fig. 3-19   Block diagram of solution to Prob. 10.

b. Let's check our network. The T feedback network can be changed to a delta as described in Appendix D1. The resulting circuit is shown in Fig. 3-20 after conversions. Z_A goes to ground and contributes nothing to the output. Z_C is directly across the output and it also has no contribution. Z_B is all we are concerned with.

$$Z_B = \frac{Z_1 Z_2 + Z_2 Z_3 + Z_1 Z_3}{Z_3} \qquad \text{(D1-5)}$$

$$Z_1 = Z_2 = R_1$$

$$Z_3 = R_2 + \frac{1}{Cs} = \frac{R_2 Cs + 1}{Cs}$$

$$Z_B = \frac{R_1^2 + 2R_1(R_2 Cs + 1)/Cs}{(R_2 Cs + 1)/Cs}$$

$$= 2R_1 + \frac{R_1^2 Cs}{R_2 Cs + 1}$$

$$= 18.12 \times 10^3 + \frac{8.12 \times 10^3 s}{0.047s + 1}$$

$$= \frac{8.97 \times 10^3 s + 18.12 \times 10^3}{0.047s + 1} = \frac{(18.12 \times 10^3)(0.495s + 1)}{0.047s + 1}$$

The transfer function of the compensator is

$$G_C = \frac{-Z_B}{R_3} = -\frac{18.12 \times 10^3 (0.495s + 1)}{854(0.047s + 1)} = -\frac{21.22(0.495s + 1)}{(0.047s + 1)}$$

which is a close check.

PROBLEM 11

GIVEN: The block diagram shown in Fig. 3-21. We want the system to have a phase margin of 60°.

FIND:

a. Using a lead compensator, find the transfer function that satisfies these requirements. The gain at dc shall remain the same as for the uncompensated case.

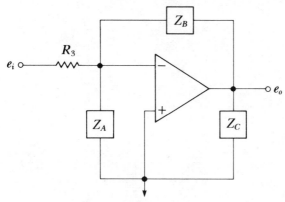

Fig. 3-20   Equivalent circuit used in the check of Prob. 10.

b. Provide a circuit arrangement that will satisfy the transfer function.

SOLUTION:

a. The procedures for the design of a lead compensator are given in many texts. First plot the magnitude and phase as shown in Fig. 3-22. For a starting point in plotting, let's choose $\omega = 0.1$ rad/s because it is far from a breakpoint and will not introduce much error.

$$GH(j\omega) = \frac{10}{j\omega(2 + j\omega)}$$

$$GH(j0.1) \approx \frac{10}{(j0.1)(2)} = -j50$$

$$|GH(j0.1)| = 34 \text{ dB}$$

From our plot, we can see that our phase ϕ_x at 0 dB gain is approximately $-146°$, which informs us our phase margin is $+34°$ $(180° - 146°)$. If we want a phase margin of $+60°$, we must introduce a phase shift of $\phi_m = $ phase margin $- 180 - \phi_x =$

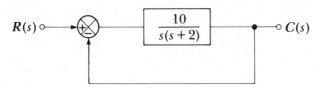

Fig. 3-21   Block diagram for Prob. 11.

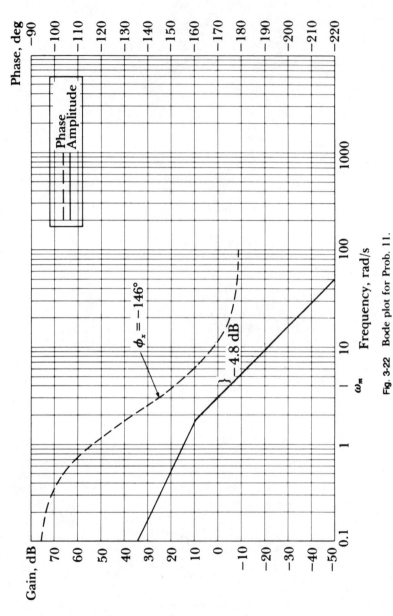

Fig. 3-22 Bode plot for Prob. 11.

152

$60 - 180 + 146 = 26°$. The normal procedure is to add a few degrees because the crossover shifts on the axis in a way to introduce more lag. Use 30°.

$$\sin \phi_m = \frac{\alpha - 1}{\alpha + 1}$$

$$\frac{\alpha - 1}{\alpha + 1} = 0.5$$

$$\alpha = 3$$

$$GH(\omega_m) = -10 \log \alpha = -10 \log 3 = -4.8 \, dB$$

If we again refer to Fig. 3-22, we find that at $GH(\omega_n) = -4.8$ dB, ω_m is approximately 4 rad/s, which will be our new 0 dB crossover. T_1 and $G_C(s)$ can now be calculated:

$$T_1 = \alpha^{-1/2}/\omega_m = 1/[(1.73)(4)] = 0.145 \text{ s}$$

and

$$G_C(s) = \frac{1 + \alpha T_1 s}{\alpha(1 + T_1 s)} = \frac{1}{3}\left(\frac{1 + 0.43s}{1 + 0.145s}\right)$$

$$G_C(s) = \frac{s + 2.3}{s + 6.9}$$

b. A network that will provide this function can be found by reviewing the Bode plots of Appendix K and selecting the appropriate circuit. This circuit will be followed by a buffer amplifier to provide isolation and gain. The gain is required to meet our dc gain requirements. The circuit of Fig. K1-7 will meet our requirements and is shown in Fig. 3-23. The transfer function we desire is

$$G_C(s) = \frac{1 + 0.43s}{3(1 + 0.145s)} = \frac{A(T_1 s + 1)}{T_3 s + 1} \qquad \text{(K1-7)}$$

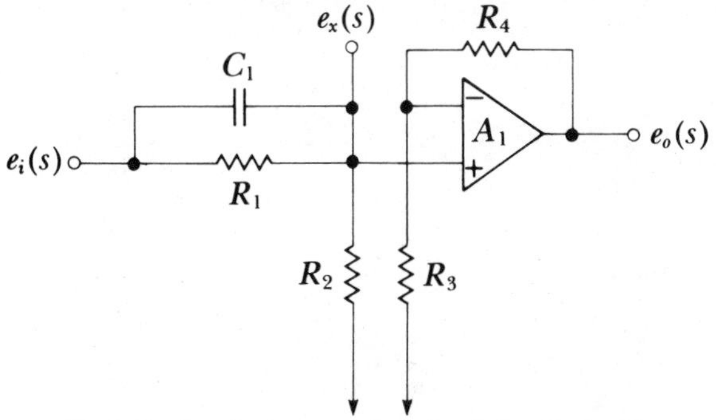

Fig. 3-23 Lead-compensation circuit for Prob. 11.

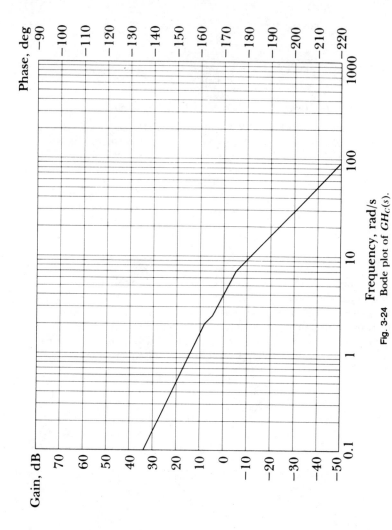

Phase, deg

Gain, dB

Frequency, rad/s
Fig. 3-24 Bode plot of $GH_C(s)$.

154

Pick $C_1 = 100\ \mu F$ and $R_3 = 10\ k\Omega$ and solve for our other values:

$$R_1 = T_1/C_1 = 0.435/10^{-4} = 4.35\ k\Omega \qquad \text{(K1-7)}$$

$$A = R_2/(R_1 + R_2) \qquad \text{(K1-7)}$$

$$R_2 = 2.15\ k\Omega$$

In order that our gain at dc remain the same, the gain of the amplifier must equal α.

$$\frac{R_4 + R_3}{R_3} = \alpha = 3 \qquad \text{(C1-1)}$$

$$R_4 = 20\ k\Omega$$

CHECK THE SOLUTION: Plot the $GH_c(s) = GH(s)G_c(s)$ on a Bode plot (see Fig. 3-24). The 0-dB crossover is found around 4 rad/s. The gain at this frequency is $GH_c(j4) = 0.97\underline{/-123.4°}$ (vs. a gain of 1 and a phase of $-120°$). This is close.

PROBLEM 12

GIVEN: The feedback control system shown in Fig. 3-25.

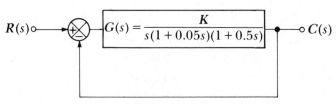

Fig. 3-25   Feedback control system for Prob. 12.

FIND:

a.   The phase-lag compensator that will fulfill the requirements of

$$K_v = 20\ s^{-1} \qquad \text{(velocity error constant)}$$

$$\phi_m = 45° \qquad \text{(phase margin)}$$

b.   A circuit that will provide us with the compensating network.

SOLUTION:

a.   The techniques for designing lag compensators are given in many texts. Let's place $GH(s)$ in the proper format first and plot. Note that $H(s) = 1$.

$$GH(s) = \frac{K}{s(1 + 0.05s)(1 + 0.5s)}$$

$$= \frac{40K}{s(s + 20)(s + 2)}$$

To meet our first requirement (that our velocity error constant be $20\ s^{-1}$), we must solve for K_v as follows:

$$K_v = \lim_{s \to 0} sG(s) = \frac{40K}{(20)(2)} = 20$$

Therefore, $K = 20$ and our transfer function becomes

$$GH(s) = \frac{800}{s(s + 2)(s + 20)}$$

$$GH(j\omega) = \frac{800}{j\omega(2 + j\omega)(20 + j\omega)}$$

Our Bode plot is shown in Fig. 3-26 for this function. From this plot we can determine that $\omega_o = 6.3$ rad/s. Let's calculate our phase at this frequency. (This is more accurate than obtaining phase from the plot.)

$$GH(j6.3) = \frac{800}{j6.3(2 + j6.3)(20 + j6.3)}$$

$$= \frac{800}{(6.3\underline{/90°})(6.61\underline{/72.4°})(20.97\underline{/17.5°})} = 0.916\underline{/-179.9°}$$

Our gain is close to 0 dB (−0.8 dB) and our phase checks. We have 0° phase margin. We need to obtain a phase of at least −135° to obtain a phase margin of 45°. Let's add an additional 5° for a safety margin. Our ϕ_c is now −130°. This occurs at a frequency ω_c of approximately 1.5 rad/s. Our gain G_c at this frequency is 22 dB. In order to cross 0 dB at 1.5 rad/s, our lag compensator must reduce the gain by this amount. The transfer function of a lag compensator is represented by

$$G_c(s) = \frac{1 + \alpha T_1 s}{1 + T_1 s}$$

The constants can be determined as follows:

$$\alpha = 10^{-G_c/20} = 10^{-22/20} = 0.0794$$
$$\alpha T_1 = 10/\omega_c = 10/1.5 = 6.67\ s$$
$$T_1 = 84\ s$$

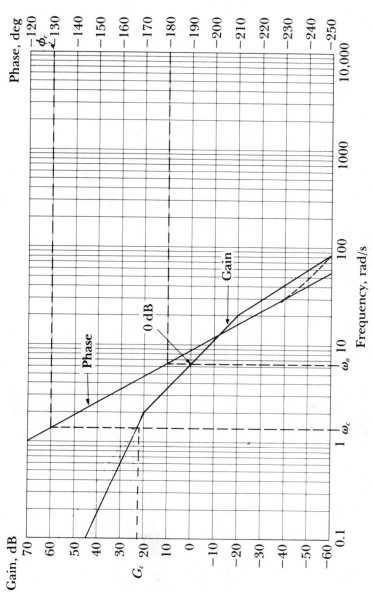

Fig. 3-26 Bode plot for $GH(s)$ of Prob. 12.

157

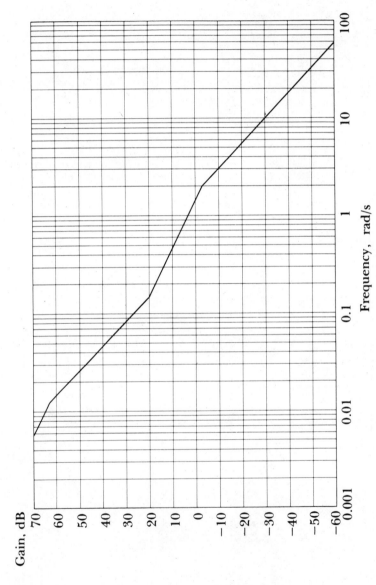

Gain, dB

Frequency, rad/s

Fig. 3-27 Bode plot of $GH_C(s)$ for Prob. 12.

158

Our transfer function becomes

$$G_c(s) = \frac{1 + 6.67s}{1 + 84s}$$

Our total compensated loop transfer function is now:

$$GH_c(s) = GH(s)G_c(s) = \frac{63.5(s + 0.015)}{s(s + 0.012)(s + 2)(s + 20)}$$

$$GH_c(j\omega) = \frac{63.5(0.15 + j\omega)}{j\omega(0.012 + j\omega)(2 + j\omega)(20 + j\omega)}$$

This function is plotted in Fig. 3-27.

b. The transfer functions of Appendix K can be reviewed, and a circuit selected to meet our $G_c(s)$ transfer function. The circuit of (K1-5) meets our requirements. Follow this circuit with a buffer amplifier to prevent loading. From Table (K1-5),

$$G_c(s) = \frac{1 + 6.67s}{1 + 84s} = \frac{T_1 s + 1}{T_3 s + 1} \qquad \text{(K1-5)}$$

$$T_1 = R_1 C_1 = 6.67 \qquad \text{(K1-5)}$$

$$T_3 = (R_1 + R_2)C_1 = 84 \qquad \text{(K1-5)}$$

Pick $C_1 = 150\ \mu\text{F}$ and solve for the remaining component values:

$$R_1 = 44.5\ \text{k}\Omega$$
$$R_2 = 516\ \text{k}\Omega$$

We would pick the nearest standard values from Appendix H in actual practice. The circuit is shown in Fig. 3-28.

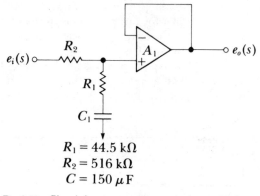

$$R_1 = 44.5\ \text{k}\Omega$$
$$R_2 = 516\ \text{k}\Omega$$
$$C = 150\ \mu\text{F}$$

Fig. 3-28   Circuit lag-compensator solution to Prob. 12.

CHECK THE SOLUTION: Check $GH(j\omega)$ at the 1.5 rad/s crossover frequency. $GH(j1.5) = 0.851\underline{/-137.2°}$. The phase margin is 42.8°. At 1.4 rad/s $GH(j1.4) = 0.936\underline{/-135.4°}$. Also the transfer function can be checked:

$$\frac{e_o(s)}{e_{in}(s)} = \frac{6.675s + 1}{84.08s + 1} \left(vs. \frac{6.67s + 1}{84s + 1} \right)$$

PROBLEM 13

GIVEN: The feedback system shown in Fig. 3-29.

FIND:

a.   Using the Nichols chart, find the gain and phase margins and at what frequency they occur.

b.   The maximum closed-loop gain M_p and 3-dB bandwidth ω, and the frequencies at which they occur.

c.   How much would we have to change the gain to achieve a phase margin of 45°?

SOLUTION:

a.   The loop transfer function in terms of ω can be expressed as:

$$GH(j\omega) = \frac{4}{j\omega(1 + j\omega)(3 + j\omega)}$$

The first step is to plot the function on a Nichols chart from the data calculated in Table 3-1. The coordinates of the Nichols chart define the open loop characteristics with the magnitude (in dB) on the ordinate and the phase (in degrees) on the abscissa. The circular contours define the closed-loop gain and phase for the function $G/(1 + G)$. The Nichols plot for this problem is shown in Fig. 3-30. A review of Appendix G1-6 provides an indication of the plot to be expected for this problem and illustrates how to find the gain and phase margins. The open-loop phase margin is defined by the 0-dB crossing of the plot defined as point A on the chart. The phase is approximately $-148°$ at this point and the phase margin is $\phi_m =$

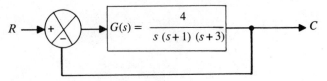

$$G(s) = \frac{4}{s\,(s+1)\,(s+3)}$$

Fig. 3-29   Feedback system for Prob. 13.

TABLE 3-1   Table for $G(j\omega) = \dfrac{4}{j\omega(1 + j\omega)(3 + j\omega)}$

ω	$\dfrac{1}{j\omega}$	$\dfrac{1}{1 + j\omega}$	$\dfrac{4}{3 + j\omega}$	$|G(j\omega)|$	$\underline{/G(j\omega)}$	$\dfrac{|G(j\omega)|}{(dB)}$
0.2	5.0 $\underline{/-90°}$	0.98$\underline{/-11.3°}$	1.32$\underline{/-3.8°}$	6.47	$-105.1°$	16.2
0.4	2.5 $\underline{/-90°}$	0.93$\underline{/-21.8°}$	1.32$\underline{/-7.6°}$	3.07	$-119.4°$	9.7
0.6	1.67$\underline{/-90°}$	0.86$\underline{/-31.0°}$	1.32$\underline{/-11.3°}$	1.90	$-132.3°$	5.56
0.8	1.25$\underline{/-90°}$	0.78$\underline{/-38.7°}$	1.28$\underline{/-14.9°}$	1.25	$-143.6°$	1.9
1.0	1.0 $\underline{/-90°}$	0.71$\underline{/-45.0°}$	1.28$\underline{/-18.4°}$	0.91	$-153.4°$	-0.83
1.2	0.83$\underline{/-90°}$	0.64$\underline{/-50.2°}$	1.24$\underline{/-21.8°}$	0.66	$-162.0°$	-3.6
1.4	0.71$\underline{/-90°}$	0.58$\underline{/-54.5°}$	1.20$\underline{/-25°}$	0.49	$-169.5°$	-6.1
1.6	0.63$\underline{/-90°}$	0.53$\underline{/-60.9°}$	1.16$\underline{/-28°}$	0.39	$-176.0°$	-8.18
1.8	0.56$\underline{/-90°}$	0.49$\underline{/-61.0°}$	1.14$\underline{/-31°}$	0.31	$-182.0°$	-10.17

$180 - 148 = 32°$. The gain margin is determined by the intercept of our plot with the 180° ordinate (point B), which is approximately -9.5 dB. From Table 3-1 we can see that a phase of $-148°$ occurs midway between the frequencies of 0.8 and 1.0 rad/s or 0.9 rad/s; and for our gain margin, we see that $\omega \approx 1.8$ rad/s at $-182°$ of phase shift.

b.   The maximum closed-loop gain occurs where the plot is tangential to the closed-loop gain contours or $M_p = 6$ dB at a frequency of 1 rad/s. To find the closed-loop phase, we go to the -3-dB contour and find that the closed-loop phase is $-180°$ at a frequency of approximately $\omega_r = 1.5$ rad/s. Check the closed-loop gain G_C at 1.0 and 1.5 rad/s to verify our observations.

$$G(j\omega) = \frac{4}{j\omega(1 + j\omega)(3 + j\omega)} = \frac{N}{D}$$

$$G_C(j\omega) = \frac{G}{1 + G} = \frac{N/D}{1 + N/D} = \frac{N}{N + D}$$

$$= \frac{4}{4 + j\omega(1 + j\omega)(3 + j\omega)}$$

$$G_C(j1.0) = \frac{4}{4 + j1(1 + j1)(3 + j1)}$$

$$G_C(j1.0) = \frac{4}{4 + (1\underline{/90°})(1.41\underline{/45°})(3.16\underline{/18.4°})}$$

$$= \frac{4}{4 + 4.69\underline{/153.4°}} = \frac{4}{4 - 4.19 + j2.1} = \frac{4}{2.11\underline{/95.3°}}$$

$$= 1.9\underline{/-95.3°} \quad \text{or} \quad 5.58 \text{ dB } \underline{/-95.3°}$$

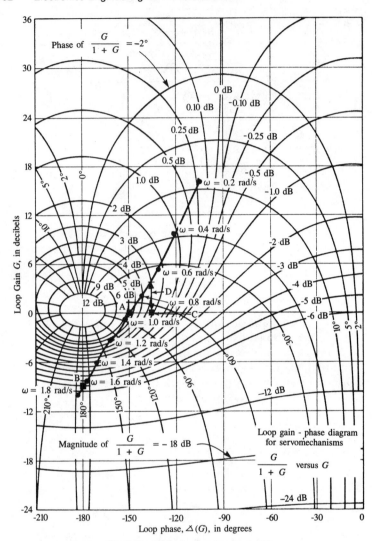

Fig. 3-30 Nichols plot for Prob. 13.

This is close to the 6 dB expected. 6 dB is a gain of 1.78. Now check our closed-loop phase at 1.5 rad/s.

$$G_C(j1.5) = \frac{4}{4 + (j1.5)(1 + j1.5)(3 + j1.5)}$$

$$= \frac{4}{4 + (1.5\underline{/90°})(1.8\underline{/56.3°})(3.35\underline{/26.6°})}$$

$$G_C(j1.5) = \frac{4}{4 + 9.045\underline{/177.3°}} = \frac{4}{4 - 9.03 + j0.426}$$

$$G_C(j1.5) = \frac{4}{-4.03 + j0.426} = \frac{4}{4.05\underline{/17.4°}}$$

$$= 0.987\underline{/-174°} \quad \text{or} \quad -0.11 \text{ dB}\underline{/-174°}$$

which is close to 0 dB at $-180°$.

c. For a phase margin of 45° we need a phase of $-135°$ for $G(j\omega)$. This phase occurs at point C on the Nichols chart and is approximately 4 dB less than the plotted $G(j\omega)$ and defined by the distance D on the chart. This occurs at 0.7 rad/s. Check $G(j\omega)$ at this frequency with the gain reduced by 4 dB (0.631). Define this as $G'(j0.7)$.

$$G'(j0.7) = \frac{0.631(4)}{(j0.7)(1 + j0.7)(3 + j0.7)}$$

$$= \frac{2.52}{(0.7\underline{/90°})(1.22\underline{/35°})(3.08\underline{/13.1°})}$$

$$G'(j0.7) = 0.961\underline{/138.1°}$$

which yields -0.34 dB and a phase margin of 41.9°. We could get the exact answer by an iterative process, but this is close enough for an examination-type problem.

CHECK THE SOLUTION: We've checked each part as we've proceeded. However, a closer look at M_p reveals that M_p occurs at $\omega = 0.95$. At this frequency, $G_C(j0.95) = 5.9$ dB $\underline{/-79°}$.

PROBLEM 14 _____

GIVEN: The plant shown in Fig. 3-31.

FIND: It is desired to find the value of K for which the system becomes unstable. Use the root-locus approach to determine K and draw the diagram.

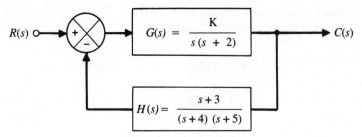

Fig. 3-31   Plant for Prob. 14.

SOLUTION: We can construct the root locus by starting with the characteristic equation:

$$1 + G(s)H(s) = 0$$

where $\quad G(s) = \dfrac{K}{s(s + 2)} \quad$ and $\quad H(s) = \dfrac{s + 3}{(s + 4)(s + 5)}$

Our characteristic equation is now:

$$1 + \frac{K(s + 3)}{s(s + 2)(s + 4)(s + 5)} = 0$$

The locations of the poles in the s plane of Fig. 3-32 are shown as X and the zero as O. Let p represent the number of poles and z the number of zeros. The number of poles is always greater than the number of zeros for all real systems and the root loci for $K = 0$ start at the open-loop poles and terminate at an open-loop zero or at infinity. For

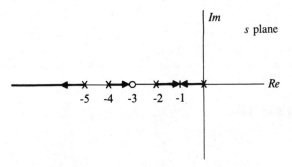

Fig. 3-32   Real axis root loci of Prob. 14.

this problem, one branch terminates at a zero and three terminate at ∞. The root loci can be shown on the real axis in Fig. 3-33. A point in the real axis falls on the loci if the number of open-loop poles and zeros to the right of the point is odd.

The asymptotes of the root loci can be determined from

$$\theta = \frac{(+/-\ 180)(2N - 1)}{p - z} \qquad N = [0, 1, 2, \ldots, (p - z - 1)]$$

$$p = 4 \qquad z = 1 \qquad N = 4 - 1 - 1 = 2$$

$$\theta = \frac{+/-\ 180(0 + 1)}{3}, \frac{+/-\ 180(2 + 1)}{3}, \frac{+/-\ 180(4 + 1)}{3} \text{ degrees}$$

$$\theta = +/-\ 60°, +/-\ 180°, +/-\ 300°$$

Of course, $+/-\ 300°$ is $+/-\ 60°$. Let p_x be the numerical value of the pole interrupts on the real axis, and z_x, the numerical value of the

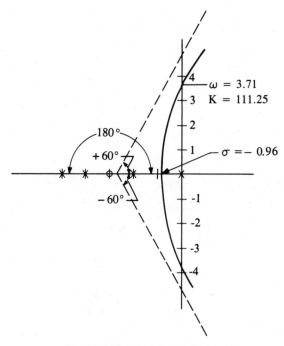

Fig. 3-33   Root loci for Prob. 14.

zero. The asymptotes intersect the real axis at the point defined by

$$\sigma = \frac{\Sigma p_x - \Sigma z_x}{p - z} = \frac{(0 - 2 - 4 - 5) - (-3)}{3}$$

$$\sigma = -8/3 = -2.67$$

The asymptotes are shown in Fig. 3-33 as the dashed lines.

The next step is to find the break-away and break-in points on the real axis. If the root locus lies between two poles on the real axis, there will be at least one break-away point. Similarly, if the root locus lies between two adjacent zeros on the root locus, there will be at least one break-in point; and for a root locus between a zero and a pole there will not be either a break-in or a break-away point. The break-in or break-away points can be found from the characteristic equation:

$$1 + \frac{KN(s)}{D(s)} = 0$$

The break-in and break-away points can be found by solving:

$$D'(s)N(s) - D(s)N'(s) = 0$$

where the primes denote differentiation with respect to s.

For this problem:

$$1 + \frac{KN(s)}{D(s)} = \frac{K(s + 3)}{s(s + 2)(s + 4)(s + 5)}$$

Therefore:

$$N(s) = s + 3 \qquad N'(s) = 1$$

$$D(s) = s(s + 2)(s + 4)(s + 5) = s^4 + 11s^3 + 38s^2 + 40s$$

$$D'(s) = 4s^3 + 33s^2 + 76s + 40$$

Substituting into our previous equation:

$$(4s^3 + 33s^2 + 76s + 40)(s + 3) - (s^4 + 11s^3 + 38s^2 + 40s) = 0$$

Reducing yields:

$$3s^4 + 34s^3 + 137s^2 + 228s + 120 = 0$$

We expect the break-away point to lie between 0 and -2. An iterative process yields the following as a solution (substitute values into the equation until it equals zero). We find:

$$s = -0.96$$

Our next step is to determine where the root locus crosses the imaginary axis. Let $s = j\omega$ in the characteristic equation and set both the real and imaginary parts equal to zero and solve for ω and K.

$$1 + \frac{K(j\omega + 3)}{j\omega(j\omega + 2)(j\omega + 4)(j\omega + 5)} = 0$$

Simplifying and combining terms:

$$(\omega^4 - 38\omega^2 - 3K) + 2(K\omega + 40\omega - 11\omega^3) = 0$$

Working with the imaginary terms:

$$2(K\omega + 40\omega - 11\omega^3) = 0$$
$$K = 11\omega^2 - 40$$

Substitute these into the expression and set equal to zero:

$$\omega^4 - 38\omega^2 + 33\omega^2 - 120 = 0$$
$$\omega^4 - 5\omega^2 - 120 = 0$$

Letting $\omega^2 = x$, we have:

$$x^2 - 5x - 120 = 0$$
$$x = \frac{5 +/- (25 + 480)^{1/2}}{2}$$
$$x = \frac{5 +/- 22.5}{2} = -8.75, 13.75$$
$$\omega^2 = 13.75$$
$$\omega = 3.71$$
$$K = 11\omega^2 - 40$$
$$K = 111.25$$

The root loci are shown in Fig. 3-33.

CHECK THE SOLUTION: Use Routh-Hurwitz and expand as a function of K.

$$1 + G(s)H(s) = 1 + \frac{K(s + 3)}{s(s + 2)(s + 4)(s + 5)} = 0$$
$$s^4 + 11s^3 + 38s^2 + (40 + K)s + 3K = 0$$

The Routh Array is

s^4	1	38	$40 + K$
s^3	11	$40 + K$	0
s^2	$\dfrac{378 - K}{11}$	$3K$	0
s^1	$\dfrac{\dfrac{(378 - K)(40 + K)}{11} - 33K}{\dfrac{378 - K)}{11}}$	0	0
s^0	$3K$	0	0

Reduce the s^1 term, set equal to zero, and solve for K.

$$40 + K - \frac{363K}{378 - K} = 0$$

$$K^2 + 25K - 15{,}120 = 0$$

$$K = \frac{-25 +/- (625 + 60{,}480)^{1/2}}{2}$$

$$K = \frac{-25 +/- 247.2}{2} = 111.1, \; -136.1$$

And 111.1 is close to the value of $K = 111.25$ solved for before. This checks our solution.

PROBLEM 15

GIVEN: The plant shown in Fig. 3-34.

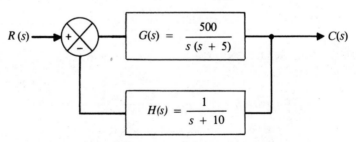

Fig. 3-34   Plant of Prob. 15.

FIND:

a. Using the Nyquist criterion, the gain and phase margins.

b. What could we do to change our phase margin to 30°?

SOLUTION:

a. The open-loop transfer function is given by:

$$GH(s) = G(s)H(s) = \frac{500}{s(s + 5)(s + 10)}$$

and as a function of ω:

$$GH(j\omega) = \frac{500}{j\omega(j\omega + 5)(j\omega + 10)}$$

We can get an idea what the polar plot looks like by reviewing Appendix G-6. If we encircle the $\sigma = -1$ point on the real axis, the system is unstable. The calculations for the polar plot of Fig. 3-35 are provided in Table 3-2. An expanded version of Fig. 3-35 is provided in Fig. 3-36 where the unity gain circle is much larger. A review of Fig. 3-35 demonstrates that the plot does not enclose the point -1 and is stable. The phase margin is determined from the point where the plot intersects the unity gain circle. From Table 3-2 and Fig. 3-36, we can determine that this occurs at approximately

TABLE 3-2 $GH(j\omega)$ gain and phase calculation for Prob. 15

ω	$\dfrac{1}{5 + j\omega}$	$\dfrac{1}{10 + j\omega}$	$\dfrac{500}{j\omega}$	$GH(j\omega)$
1	$0.196\angle-11.3°$	$0.099\angle-5.7°$	$500\angle-90°$	$9.7\angle-107.0°$
2	$0.185\angle-21.8°$	$0.098\angle-11.3°$	$250\angle-90°$	$4.5\angle-123.1°$
3	$0.172\angle-31.0°$	$0.096\angle-16.7°$	$167\angle-90°$	$2.8\angle-137.7°$
4	$0.156\angle-36.7°$	$0.093\angle-21.8°$	$125\angle-90°$	$1.8\angle-148.5°$
5	$0.141\angle-45.0°$	$0.089\angle-26.6°$	$100\angle-90°$	$1.25\angle-161.6°$
6	$0.128\angle-50.0°$	$0.085\angle-31.0°$	$83.3\angle-90°$	$0.91\angle-171.0°$
7	$0.116\angle-54.5°$	$0.082\angle-35.0°$	$71.4\angle-90°$	$0.68\angle-180.0°$
8	$0.106\angle-58.0°$	$0.080\angle-38.7°$	$62.5\angle-90°$	$0.52\angle-186.7°$
9	$0.097\angle-60.9°$	$0.074\angle-42.0°$	$55.6\angle-90°$	$0.40\angle-192.9°$
10	$0.089\angle-63.4°$	$0.071\angle-45.0°$	$50.0\angle-90°$	$0.32\angle-198.4°$
15	$0.063\angle-71.6°$	$0.056\angle-56.3°$	$33.3\angle-90°$	$0.12\angle-217.9°$

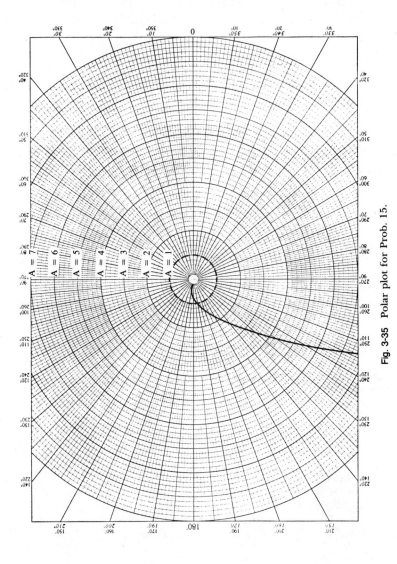

Fig. 3-35 Polar plot for Prob. 15.

170

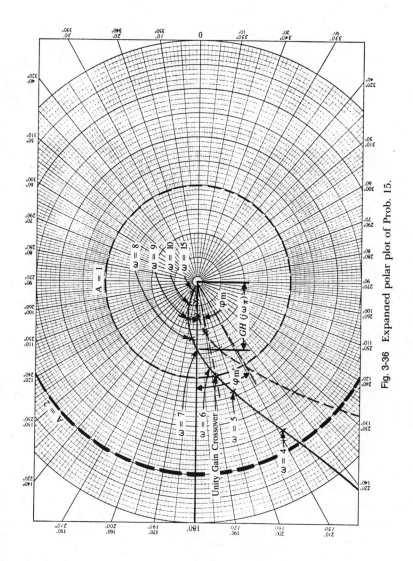

Fig. 3-36 Expanded polar plot of Prob. 15.

171

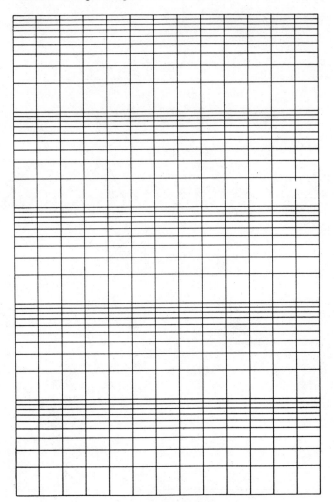

Fig. 3-37   Bode plot.

−168° and ω can be interpolated to be 5.7 rad/s. Our phase margin is

$$\phi m = 180 - 168 = 12°$$

We would normally prefer +45°. The gain margin can be determined from the plot also. The gain margin can be determined from

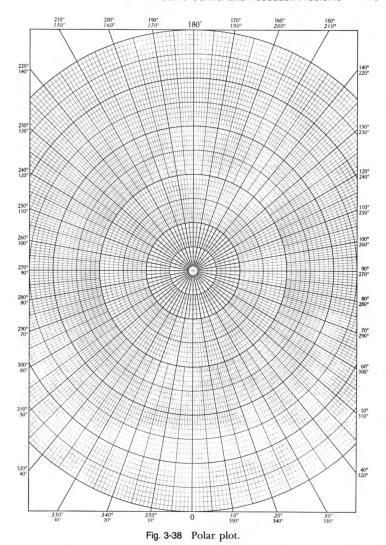

Fig. 3-38 Polar plot.

Fig. 3-36 where the plot crosses 180°. This is labeled $GH(j\omega_\pi)$ and occurs at $\omega = 7$ rad/s. Our gain margin in dB is

$$GM = 20 \log \left(\frac{1}{|GHj\omega_\pi|} \right) = 3.35 \text{ dB}$$

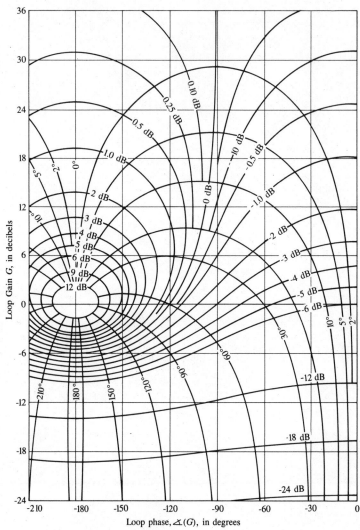

Fig. 3-39   Nichols chart.

b.   The obvious solution is to reduce gain at the expense of band-width and sensitivity. And if these two features are important, then we could use a lead-lag type of compensation. Let's assume that we can suffer the consequences associated with a reduction in gain. Our gain at $\omega = 5.8$ rad/s is approximately $1 \angle -168°$. For 30° of phase margin we need an angle of -150. This occurs at $\omega \approx 4$ rad/s. Our gain is 1.8 at this frequency. Therefore, we only need to reduce our gain from 1.8 to 1, a factor of 44%. Thus our transfer function becomes

$$GH(s)' = \frac{278}{s(s + 5)(s + 10)}$$

And the plot is shown as the dotted line in Fig. 3-36. The new phase margin is shown as $\phi m'$.

CHECK THE SOLUTION: The best check would be to check the phase at $\omega = 5.7$ rad/s for the original transfer function. Then we should check the new transfer function at $\omega = 4$ rad/s.

$$GH(j5.8) = \frac{500}{j5.7(j5.7 + 5)(j5.7 + 10)}$$

$$= \frac{500}{(5.7\angle 90°)(7.6\angle 48.7°)(11.5\angle 29.7°)} = \angle -168.3°$$

which checks very closely with unity gain at $-168°$ estimated previously.

$$GH(j4)' = \frac{278}{j4(j4 + 5)(j4 + 10)} = \frac{278}{(4\angle 90°)(6.4\angle 38.7°)(10.8\angle 21.8°)}$$

$$= 1\angle -150.5°$$

This checks our phase margin the 29.5°, which is close to the 30° designed for.

BLANK BODE, POLAR, AND NICHOLS CHARTS

The charts in Figs. 3-37, 3-38, 3-39 are for use in working out the problems found on the test.

4

INSTRUMENTATION, COMPUTER, AND SYSTEM PROBLEMS

The subjects covered in this chapter include instrumentation, computer, and system problems. Appendices C, D, E, and H to L are the most useful in the solution of problems of the type presented in this chapter.

PROBLEM 1

GIVEN: The amplifier shown in Fig. 4-1, used to measure temperature. R_T is a temperature-sensitive resistor that varies with temperature as follows: $R_T = 1000e^{-T/25°C}$

FIND:

a. R_2, if E_o is to be 0 V at $-55°C$.

b. R_5, if we want a full-scale deflection at 125°C. The meter resistance R_M is 1 kΩ and the meter has a full-scale deflection of 1 mA.

SOLUTION: First find the values of R_T at $-55°C$ and 125°C.

At $-55°C$: $\qquad R_T = 1000e^{-55/-25} = 9.025$ kΩ

At 125°C: $\qquad R_T = 1000e^{-125/25} = 6.74$ Ω

From Appendix C3 and Fig. C3-1, we have the following equivalencies:

Figure C3-1	Figure 4-1	Value
R_1	R_3	10 kΩ
R_2	R_4	10 kΩ
R_3	R_1	10 kΩ
R_4	$R_2 + R_T$	
V_1	15 V dc	15 V dc
V_2	0	0
V_3	15 V dc	15 V dc

Rewrite Eq. (C3-1):

$$E_o = \frac{(V_1 - V_2)(R_3 + R_4)R_2}{(R_1 + R_2)R_3} + \frac{(R_3 + R_4)V_2}{R_3} - \frac{V_3 R_4}{R_3} \qquad \text{(C3-1)}$$

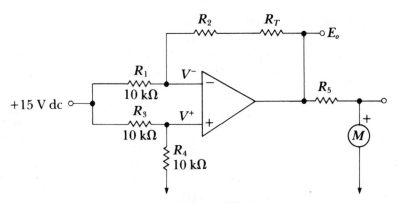

Fig. 4-1 Amplifier of Prob. 1.

Substitute the terms for this problem and let $V = 15$ V dc:

$$E_o = \frac{V(R_1 + R_2 + R_T)R_4}{(R_3 + R_4)R_1} - \frac{V(R_2 + R_T)}{R_1}$$

Solve for R_2 with $E_o = 0$ V and $R_T = 9.025$ kΩ.

$$0 = \frac{15(10 \times 10^3 + R_2 + 9.025 \times 10^3)10 \times 10^3}{(20 \times 10^3)10 \times 10^3} - \frac{15(R_2 + 9.025 \times 10^3)}{10 \times 10^3}$$

Solving, we find

$$0.5R_2 + 9.5125 \times 10^3 - R_2 - 9.025 \times 10^3 = 0$$

$$R_2 = 975 \ \Omega$$

From Appendix H1, select $R_2 = 976 \ \Omega$.

When we are at maximum temperature, we want full-scale deflection. Calculate the value of E_o at 125°C ($R_T = 6.74 \ \Omega$ from before):

$$E_o = \frac{15(10 \times 10^3 + 975 + 6.74)10 \times 10^3}{(20 \times 10^3)(10 \times 10^3)} - \frac{15(975 + 6.74)}{10 \times 10^3}$$

$$= 6.764 \text{ V dc}$$

With this value of E_o, we want 1 mA to flow into the meter. Therefore,

$$R_5 + R_M = \frac{E_o}{1 \text{ mA}} = \frac{6.764}{1 \text{ mA}} = 6.764 \text{ k}\Omega$$

$$R_5 = 5.764 \text{ k}\Omega$$

Choose the nearest 1% value; let $R_5 = 5.76 \text{ k}\Omega$.

CHECK THE SOLUTION: Solve for E_o at -55°C: $V^+ = V^- = 7.5$ V dc, $I_{R1} = I_{R2} = 0.75$ mA, and $E_o = V^- - I_R(R_2 + R_T) = 0$ V. At 125°C, $E_o = 6.764$ V dc. The I_m meter is $6.764/(R_m + R_5) = 1$ mA.

PROBLEM 2

GIVEN: The circuit of Fig. 4-2 with a meter whose full-scale deflection is 1 mA. The internal resistance of the meter is 10 kΩ.

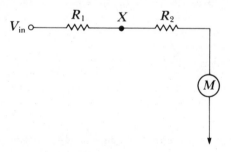

Fig. 4-2   Meter circuit of Prob. 2.

FIND:

a. What value of R_1 and R_2 would make the meter deflect full scale when 200 V is applied to the input?

b. It is desired to protect the meter by inserting a 50-V zener diode from point X to ground. What values should R_1 and R_2 have?

c. What is the ohms/volt rating of this circuit?

SOLUTION:

a. $R_1 + R_2 + R_M = \dfrac{200\ V}{1\ mA} = 200\ k\Omega \qquad R_m = \text{meter resistance}$

$R_1 + R_2 = 190\ k\Omega$

b. $\dfrac{R_2 + R_m}{R_1 + R_2 + R_m} = \dfrac{50\ V}{200\ V}$

$R_2 + 10 \times 10^3 = 0.25(190 \times 10^3) + 0.25(10 \times 10^3)$

$R_2 = 40\ k\Omega \qquad \text{(use closest standard value in}$
Appendix H1)

$R_1 = 150\ k\Omega$

c. $\text{ohms/volt} = \dfrac{200\ k\Omega}{200\ V} = 1\ k\Omega/V$

CHECK THE SOLUTION: The solution is best checked by recalculating each step.

PROBLEM 3

GIVEN: The schematic of a measuring system as shown in Fig. 4-3. The motor drives a recording stylus which indicates temperature and simultaneously moves the wiper of the potentiometer that indicates the position of the stylus. R_2 and R_3 are internal to the recorder. R_T is the transducer and has a resistance value of 330 Ω at 25°C and a temperature coefficient of $+3\ \Omega/°C$.

FIND: The value of R_1 and R_4 if we want to measure a temperature range of -25 to 125°C, and the wiper of the potentiometer is returned to ground potential. At -25°C, the stylus is fully counterclockwise (CCW).

SOLUTION: First determine the values of R_T at -25°C and $+125$°C:

At -25°C: $R_T' = 330 + (3\ \Omega/°C)(-50°C) = 180\ \Omega$
At 125°C: $R_T'' = 330 + (3\ \Omega/°C)(100°C) = 630\ \Omega$

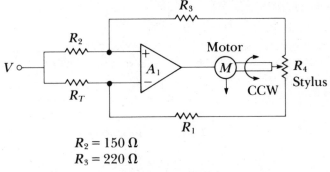

$$R_2 = 150 \ \Omega$$
$$R_3 = 220 \ \Omega$$

Fig. 4-3   Measuring system schematic of Prob. 3.

Let's draw the equivalent circuit for the −25°C condition as shown in Fig. 4-4. The potentiometer is full CCW. Since the voltage at the (−) terminal must equal the voltage at the (+) terminal, the following equations apply:

$$V^+ = V^-$$

$$\frac{(R_3 + R_4)V}{R_2 + R_3 + R_4} = \frac{R_1 V}{R'_T + R_1}$$

$$\frac{220 + R_4}{150 + 220 + R_4} = \frac{R_1}{180 + R_1}$$

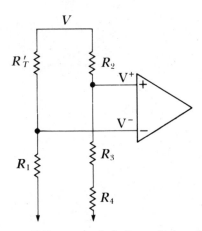

Fig. 4-4   Fully counterclockwise equivalent circuit.

Solving, we find

$$R_4 = 0.833R_1 - 220$$

For the 125°C condition, the potentiometer is fully clockwise, and the circuit of Fig. 4-5 is applicable. The following equations can be written for the 125°C condition.

$$V^+ = V^-$$

$$\frac{(R_1 + R_4)V}{R_T'' + R_1 + R_4} = \frac{R_3 V}{R_2 + R_3}$$

$$\frac{R_1 + R_4}{630 + R_1 + R_4} = \frac{220}{370}$$

Solving, we find

$$R_1 + R_4 = 925 \ \Omega$$

and substituting from our previous solution for R_4, we find

$$R_1 + (0.833R_1 - 220) = 925 \ \Omega$$

$$R_1 = 625 \ \Omega$$

$$R_4 = 300 \ \Omega$$

We would choose the nearest standard value for R_1 from Appendix H1: $R_1 = 619$; R_4 would be a potentiometer of 300 Ω.

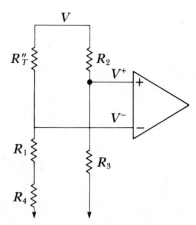

Fig. 4-5   Fully clockwise equivalent circuit.

CHECK THE SOLUTION: At $-25°C$, $V^+ = (220 + 300)V/(150 + 220 + 300) = 0.776$ V. $V^- = 625/(180 + 625) = 0.776$, which checks. AT $125°C$, $V^+ = V^- = 0.595$ V dc.

PROBLEM 4

GIVEN: A dc ammeter is used to measure the rms values of a waveform as shown in Fig. 4-6.

FIND: Calculate the value R_1 for the circuit shown in Fig. 4-7 if the internal resistance of the meter is 1 kΩ and the meter has a full-scale deflection of 1 mA. The peak input signal level is 10 V.

SOLUTION: A dc ammeter reads the average value of a waveform. Since we need rms values, we must find the rms and average values for the waveform and design the resistor accordingly. We must make the assumption that the natural time constant of the meter movement is much larger than the period of the waveform, or the meter movement will follow the profile of the waveform. Let's calculate the average value for the waveform first. The equation for the waveform from the origin to $T/2$ is $E_1(t) = (Et)/(T/2) = 2Et/T$. The equation for the waveform from $T/2$ to T is $E_2(t) = 2E - 2Et/T$. The average value can be calculated as follows:

$$E_{AV} = \frac{1}{T}\int_0^T e\, dt = \frac{1}{T}\left[\int_0^{T/2}\frac{2Et}{T}\,dt + \int_{T/2}^T\left(2E - \frac{2Et}{T}\right)dt\right]$$

$$= \frac{1}{T}\left(\frac{Et^2}{T}\Big|_0^{T/2} + 2Et - \frac{Et^2}{T}\Big|_{T/2}^T\right)$$

$$= \frac{1}{T}\left(\frac{ET}{4} + 2ET - ET - ET + \frac{ET}{4}\right) = \frac{E}{2} = 5\text{ V}$$

The rms value of the triangular waveform can be calculated as follows:

$$E_{rms} = \left(\frac{1}{T}\int_0^T e^2\,dt\right)^{1/2}$$

$$= \left[\frac{1}{T}\int_0^{T/2}\left(\frac{2Et}{T}\right)^2 dt + \frac{1}{T}\int_{T/2}^T\left(2E - \frac{2Et}{T}\right)^2 dt\right]^{1/2}$$

$$= \left[\frac{1}{T}\left(\frac{4E^2t^3}{3T^2}\right)_0^{T/2} + \frac{1}{T}\left(4E^2t - \frac{4E^2t^2}{T} + \frac{4E^2t^3}{3T^2}\right)_{T/2}^T\right]^{1/2}$$

$$= \left[\frac{1}{T}\left(\frac{E^2T}{6}\right) + \frac{1}{T}\left(4E^2T - 2E^2T - 4E^2T + E^2T + \frac{4E^2T}{3} - \frac{E^2T}{6}\right)\right]^{1/2}$$

$$= \left(\frac{E^2}{6} + 4E^2 - 2E^2 - 4E^2 + E^2 + \frac{4E^2}{3} - \frac{E^2}{6}\right)^{1/2} = \left(\frac{E^2}{3}\right)^{1/2}$$

$$= 5.774\text{ V}$$

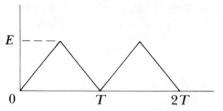

Fig. 4-6   Input waveform of Prob. 4.

For a full-scale deflection of 10 V, we would want the meter resistance plus R_1 to equal 10 kΩ for a 1 mA full-scale deflection. Thus the meter would read 5 V for the average value. But, we want an rms reading which is 5.774 V, so we need to increase our current by 5.774/5, which yields a current increase to 1.155 mA. Therefore,

$$R_M + R_1 = 10 \text{ V}/(1.155 \text{ mA}) = 8.66 \text{ k}\Omega$$

$$R_1 = 7.66 \text{ k}\Omega \quad \text{or the nearest standard value}$$

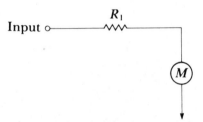

Fig. 4-7   Ammeter circuit of Prob. 4.

CHECK THE SOLUTION: The average waveform is easily checked for this waveform by computing the area and dividing by the period:

$$E_{\text{AV}} = \frac{E(T/2)\frac{1}{2}}{T} + \frac{E(T/2)\frac{1}{2}}{T} = \frac{E}{2}$$

The rms value can be checked by checking the rms value for the Fourier series components for a triangular waveform. From Appendix I, we have for the triangular waveform:

$$E(t) = \frac{E}{2} + \frac{4E \sin \omega t}{\pi^2} - \frac{4E \sin 3\omega t}{9\pi^2} + \frac{4E \sin 5\omega t}{25\pi^2} \qquad \text{(I-1)}$$

The peak value of each sinusoidal term is multiplied by 0.707 to get the rms value. The total rms value for the first four terms is

$$E_{rms} = \left\{ \left[\frac{E}{2} \right]^2 + \left[\frac{0.707(4E)}{\pi^2} \right]^2 + \left[\frac{0.707(4E)}{9\pi^2} \right]^2 + \left[\frac{0.707(4E)}{25\pi^2} \right]^2 \right\}^{1/2}$$

$$= 5.7727 \text{ V}$$

which is a close check. To be equal, we would have to take all terms of the series expansion.

PROBLEM 5

GIVEN: We are given a piece of equipment with a 20-Ω load permanently affixed across the output which measures 10.8 V dc. We are told that the equipment is a dc power supply.

FIND: A technique to determine the source voltage and equipment internal resistance.

SOLUTION: A common technique to solve a problem like this is to shunt the load with a fixed resistance and use standard circuit analysis to determine the voltage and resistance required. For best accuracy we make the shunt as small as possible, but doing this, we risk overstressing or damaging the supply. Assume a 100-Ω shunt causes the voltage to drop to 10.3 V dc. We can calculate the required values with the aid of Fig. 4-8. Let $V_{o1} = 10.8$ V dc and $V_{o2} = 10.3$ V dc, which correspond to the 20-Ω load and $20 \| 100(R_p)$-Ω load conditions, respectively.

$$V_{o1} = 10.8 \text{ V} = \frac{R_L V_s}{R_s + R_L}$$

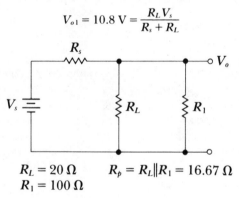

$$R_L = 20 \ \Omega \qquad R_p = R_L \| R_1 = 16.67 \ \Omega$$
$$R_1 = 100 \ \Omega$$

Fig. 4-8 Test circuit for Prob. 5.

Also

$$V_{o2} = 10.3 \text{ V} = \frac{R_p V_s}{R_s + R_p}$$

Solving, we find

$$V_s = 10.8 + 0.54R_s$$
$$V_s = 10.3 + 0.618R_s$$
$$R_s = 6.41 \ \Omega$$
$$V_s = 14.26 \ \text{V dc}$$

CHECK THE SOLUTION:

$$V_{o1} = \frac{(20)(14.26)}{26.41} = 10.8 \ \text{V dc}$$

$$V_{o2} = \frac{(16.67)(14.26)}{16.67 + 6.41} = 10.3 \ \text{V dc}$$

and our check is complete.

PROBLEM 6

GIVEN: The circuit of Fig. 4-9 is to be tested to the specified outputs of Table 4-1. The inputs are permitted to assume any value within the specified limits of Table 4-1. The output of e_b as a function of frequency is provided in Fig. 4-10. The dashed line represents the actual response, while the solid line represents the approximate response. All outputs are measured with respect to ground and the meter accuracies are $+/-$ 0.05 V ac or dc. The output error includes the variations caused by the inputs.

FIND:

a.   There are several things wrong with this test approach. What recommendations would you make to increase throughput and still maintain confidence that you are detecting faulty hardware?

b.   Assume the best meter available is accurate to $+/-$ 0.01 V ac or dc, and the best available signal generator is $+/-$ 0.01 V_{rms} instead of .05 V_{rms}. What could you do if you wanted to increase the accuracy of the total measurement of e_b to $+/-$ 0.001 V_{rms} with the available equipment?

SOLUTION:

a.   When testing a UUT (unit-under-test), the acceptable industry practice is to minimize all errors associated with the test setup such that any variations in the output are only those of the UUT. A 10:1 accuracy ratio for measuring equipment, stimuli, loads, and other ancillary apparatus is considered acceptable because the error that they introduce is minimal. For example, if our circuit output is 10

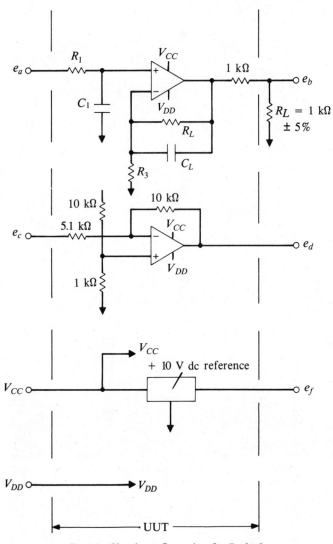

Fig. 4-9 Circuit configuration for Prob. 6.

TABLE 4-1   Calculated input/output relationships for the circuitry of Fig. 4-9

Specified Inputs	Specified Outputs
V_{CC} = 15 +/− 0.75 V dc	E_f = 10 +/− 0.05 V dc
V_{DD} = −15 +/− 0.75 V dc	E_d = 8 V +/− 0.8 V dc
e_c = −2 +/− 0.05 V dc	e_b = 10.0 +/− 1.00 at dc
e_a = 1 +/− 0.05 V_{rms}	= 7.1 +/− 0.75 at 1 kHz
	= 2.8 +/− 0.35 at 5 kHz
	= 1.7 +/− 0.25 at 10 kHz
	= 1.4 +/− 0.20 at 20 kHz
	= 1.1 +/− 0.15 at 50 kHz
	= 1.0 +/− 0.15 at 100 kHz

+/− 1 V ac (10%), our stimuli is +/− 0.1 V ac (1%), and our measuring device is +/− 0.1 V (1%), then the total output variation including the UUT is determined statistically to be

$$\Delta = [(10\%)^2 + (1\%)^2 + (1\%)^2]^{1/2} = 10.1\%$$

In other words, if our desired circuit total tolerance is +/− 10%, then the test equipment introduces +/− 0.1% absolute error or 10

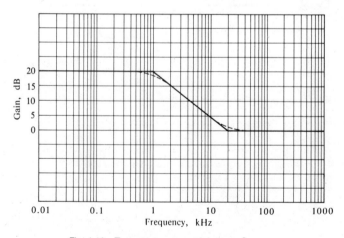

Fig. 4-10   Frequency-response curve for e_b.

mV, which is certainly acceptable. So, let's fix all of our stimuli and loads. For this problem, the following is realistic and achievable:

$V_{CC} = 15 +/- 0.01$ V dc   (instead of $+/- 0.05$ V dc)

$V_{DD} = 15 +/- 0.01$ V dc   (instead of $+/- 0.05$ V dc)

$R_L = 1 +/- 0.01$ kΩ (instead of $+/- 0.05$ kΩ)

$e_a = 1 +/- 0.01$ V$_{rms}$ (instead of $+/- 0.05$ V$_{rms}$)

$e_c = -2 +/- 0.01$ V$_{rms}$ (instead of $+/- 0.05$ V$_{rms}$)

After the stimuli are fixed, the outputs should be recalculated and specified to reflect the tighter tolerance.

E_f is specified to be $10 +/- 0.05$ V dc, yet our meter is only accurate to $+/- 0.05$ V, thus violating our 10:1 accuracy requirements. The best way to treat this violation is to test to $+/- 0.1$ V, and make a statement such as: "This reading includes an error introduced by the test equipment of 0.05 V dc." This assumes that we can tolerate $+/- 0.1$ V and eases the 10:1 accuracy requirement on the test equipment design engineer. In order to reduce tests, only measure e_b at dc and the two breakpoints, 1 kHz and 20 kHz.

b.   Since our generator is only accurate to $+/- 0.01$ V$_{rms}$ and we need $+/- 0.0001$ V$_{rms}$ to ensure a 10:1 accuracy ratio, the input should be read and recorded or stored, and the output ratioed to the input. This takes out the inaccuracy of the generator and the meter as long as they are read on the same scale. For example, if the UUT were perfect, then at dc:

$$e_b/e_a = 10/1 = 10$$

If the input increases to 1.05 V$_{rms}$, the output increases the same amount:

$$e_b/e_a = 10.5/1.05 = 10$$

Similarly, if the meter reads high or low, the same argument can be made. Ratio techniques are a common method to minimize test errors but cannot be used for absolute readings such as E_f.

Note: If there is a large difference in the input/output relationship that results in different scales for the measuring equipment, put in a precision divider so that we can still use the ratio approach efficiently.

PROBLEM 7

GIVEN: The four-stage binary counter with the clock input as shown in Fig. 4-11.

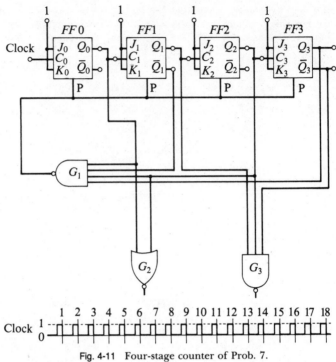

Fig. 4-11   Four-stage counter of Prob. 7.

FIND: Sketch the outputs of each flip-flop and gates G_1, G_2, and G_3 for one cycle. The function tables for the flip-flop and gates are shown in Table 4-2. Can you see anything wrong with this approach?

SOLUTION: From Fig. 4-11 or Table 4-2, gates 1 and 3 are NAND gates while gate 2 is a NOR. Gate 1 presets the counter and its output goes LO when Q_0, Q_2, and Q_3 are HI: Q_1 is LO. This corresponds to a count of 13, meaning the counter is preset right at the beginning of count 13, making it a divide by 13 counter (one state is reserved for the count zero). Gate 3 has a LO output whenever Q_1 and Q_2 are HI and Q_3 is LO. Gate 2 has a LO output whenever the outputs of Q_0 or Q_3 are HI. The waveforms are shown in Fig. 4-12.

Using a gate such as G_1 to reset more than one flip-flop is not a preferred design approach, although it is commonly used, because it depends on the flip-flop propagation delays being close enough to be preset simultaneously. A design similar to Prob. 4-9 is much preferred,

TABLE 4-2   Function tables for Prob. 7

Flip-flop Inputs				Outputs		Gates 1, 3 Inputs				Output	Gates 2 Inputs		Output
PRE	CLK	J	K	Q	$\overline{Q}$	A	B	C	D	E	A	B	C
L	X	X	X	H	L	H	H	H	H	L	L	L	H
H	↓	L	L	Q	$\overline{Q}$	All other				H	L	H	L
H	↓	H	L	L	H	combinations					H	L	L
H	↓	H	H	Toggle							H	H	L

X = don't care.

or the circuit of Fig. 4-13 may be used that ensures a preset. It provides a preset for one full clock period and is inserted between G_1 and the preset inputs.

PROBLEM 8

GIVEN: A JK flip-flop, an AND gate, and an OR gate with the function tables shown in Fig. 4-14.

FIND:

    a.   Design a synchronous counter to divide by 16.

    b.   Design a synchronous counter to divide by 14.

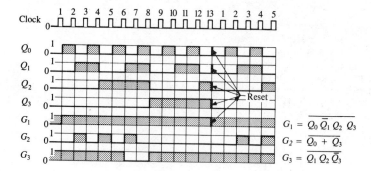

Fig. 4-12   Solution waveforms for Prob. 7.

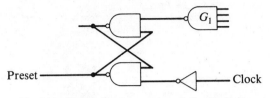

Fig. 4-13   Preset alternate approach for Prob. 7.

SOLUTION:

a. Since there are 16 states, we need 4 flip-flops ($2^4 = 16$). If we want a synchronous counter, all clock inputs must be tied together and gating used to control the J or K inputs of the subsequent stages. The steps for solving this part of the problem are as follows:

The truth table for a divide by 16 is shown in Table 4-3. From the FF function table we can see that the FF toggles when both inputs are HI. Therefore, for our first stage, we will tie both J and K a HI input. From Table 4-3, we can see that the second FF changes state only when Q_1 is a high number. Therefore, we tie the Q_1 input into the J and K inputs of Q_2. Again from Table 4-3, we see that Q_3 toggles only when Q_1 and Q_2 are at a HI level (■). Therefore, we gate Q_1 and Q_2 via gate G_1 into the J and K inputs of Q_3. Finally, Q_4 is toggled only when Q_1, Q_2, and Q_3 are all HI (*). Therefore, we

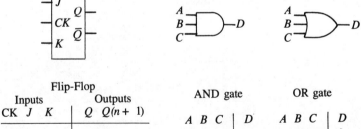

	Flip-Flop				AND gate		OR gate	
Inputs			Outputs					
CK	J	K	Q	$Q(n+1)$	A B C	D	A B C	D
↓	0	0	1	1	1 1 1	1	0 0 0	0
↓	0	0	0	0	All		All	
↓	1	0	X	1	others	0	others	1
↓	0	1	X	0				
↓	1	1	1	0				
↓	1	1	0	1				

Fig. 4-14   Function tables.

TABLE 4-3 Truth table for a divide by 16 synchronous counter

Q_n				$Q_{(n+1)}$			
Q_1	Q_2	Q_3	Q_4	Q_1	Q_2	Q_3	Q_4
0	0	0	0	1	0	0	0
1#	0	0	0	0	1#	0	0
0	1	0	0	1	1	0	0
1#	1■	0	0	0	0#	1■	0
0	0	1	0	1	0	1	0
1#	0	1	0	0	1#	1	0
0	1	1	0	1	1	1	0
1#	1■	1*	0	0	0#	0■	1*
0	0	0	1	1	0	0	1
1#	0	0	1	0	1#	0	1
0	1	0	1	1	1	0	1
1#	1■	0	1	0	0#	1■	1
0	0	1	1	1	0	1	1
1#	0	1	1	0	1#	1	1
0	1	1	1	1	1	1	1
1#	1■	1*	1	0	0#	0■	0*

TABLE 4-4 Truth table for a divide by 14 synchronous counter

Q_n				$Q_{(n+1)}$			
Q_1	Q_2	Q_3	Q_4	Q_1	Q_2	Q_3	Q_4
0	0	0	0	1	0	0	0
1	0	0	0	0	1	0	0
0	1	0	0	1	1	0	0
1	1*	0	0	0	0	1*	0
0	0	1	0	1	0	1	0
1	0	1	0	0	1	1	0
0	1	1	0	1	1	1	0
1	1*	1†	0	0	0	0*	1†
0	0	0	1	1	0	0	1
1	0	0	1	0	1	0	1
0	1	0	1	1	1	0	1
1	1*	0	1	0	0	1*	1
0	0	1	1	1	0	1	1
1	0	1	1*†	0	0	0*	0†

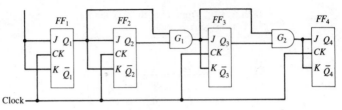

Fig. 4-15 Divide by 16 synchronous counter.

gate the output of G_2 ($G_2 = Q_1Q_2Q_3$) into the J and K inputs of Q_4. The solution is shown in Fig. 4-15.

b.   For a divide by 14 counter, set all flip-flops to zero after count 13. Don't forget count zero is a state. We can see from inspection of Table 4-4 that for Q_1 and Q_2, the J and K inputs remain the same as for the divide by 16 counter. However, for Q_3, the states change whenever the * appears. From inspection the defining equation for Q_3 is as follows:

$$Q_3(n + 1) = Q_1Q_2 + Q_1Q_3Q_4 \qquad G_3 = Q_1Q_2$$

$$G_4 = Q_1Q_3Q_4 \qquad G_5 = Q_1Q_2 + Q_1Q_3Q_4$$

Likewise for Q_4, Q_4 changes state when † appears:

$$Q_4(n + 1) = Q_1Q_2Q_3 + Q_1Q_3Q_4 \qquad G_6 = Q_1Q_2Q_3$$

$$G_4 = Q_1Q_3Q_4 \qquad G_7 = Q_1Q_2Q_3 + Q_1Q_3Q_4$$

The solution is shown in Fig. 4-16. Karnaugh maps could have been made for $Q_{3(n+1)}$ and $Q_{4(n+1)}$ to minimize the gating instead of deter-

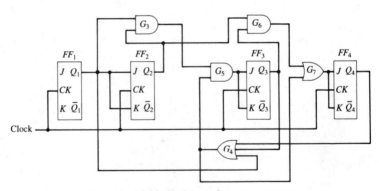

Fig. 4-16 Divide by 14 synchronous counter.

mining by inspection. On the examination Karnaugh maps are more impressive. See Prob. 4-12 for this approach.

PROBLEM 9

GIVEN: A four-bit binary-coded decimal (BCD) code as shown in Table 4-5, appearing as inputs A, B, C, and D, A being the least significant bit (LSB).

TABLE 4-5   BCD input, Gray code output for Problem 6

Input				Output			
A	B	C	D	W	X	Y	Z
0	0	0	0	0	0	0	0
1	0	0	0	1	0	0	0
0	1	0	0	1	1	0	0
1	1	0	0	1	1	1	0
0	0	1	0	1	1	1	1
1	0	1	0	0	1	1	1
0	1	1	0	0	0	1	1
1	1	1	0	0	0	0	1
0	0	0	1	1	0	0	1
1	0	0	1	1	1	0	1

FIND: A NAND gate implementation to convert from this BCD code to the Gray code shown.

SOLUTION: First we set up our standard basis which gives the designation numbers of the inputs and then those of the outputs (W, X, Y, Z) desired. This is done as shown in Table 4-6. The table is set up as follows: For the BCD portion of the table, col. 1, we have a zero code; in the bottom part of the table we also want a zero code. This procedure is continued for all columns and corresponds to the desired results shown in Table 4-5. Now, since W, X, Y, and Z are functions of A, B, C, and D, we can use a Karnaugh map to simplify each of our expressions W, X, Y, and Z. The numbers in each square correspond to the minterm representation in the designation number.

Example: $0 \to \bar{A}\bar{B}\bar{C}\bar{D}$ $2 \to \bar{A}B\bar{C}\bar{D}$

We will go through the complete procedure for determining W, and the solutions for X, Y, and Z can be done similarly. The designation number for W is determined from Table 4-6:

$$W = 0111 \quad 1000 \quad 11XX \quad XXXX$$

The X's denote constrained states or "don't care" conditions. The Karnaugh map of W is shown in Fig. 4-17. In all boxes where we want a 1 we put a diagonal line. In all squares where we "don't care" we put an X. The X means that the square can be occupied by either a 1 or a

TABLE 4-6   Standard basis for Prob. 9

A	0101	0101	01	(LSB)		
B	0011	0011	00		BCD	(Input)
C	0000	1111	00		code	
D	0000	0000	11			
W	0111	1000	11	(LSB)		
X	0011	1100	01		Gray	(Output)
Y	0001	1110	00		code	
Z	0000	1111	11			

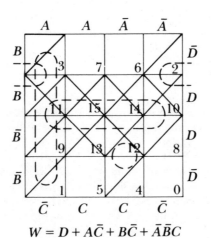

$$W = D + A\bar{C} + B\bar{C} + \bar{A}\bar{B}C$$

Fig. 4-17 Karnaugh map of W.

0 at the convenience of the designer. Now minimization can be achieved. All *X*'s are taken to be **1** for this variable. Grouping as shown yields:

Squares	Representation
3, 11, 9, 1	$A\bar{C}$
2, 3, 10, 11	$B\bar{C}$
8, 9, 10, 11, 12, 13, 14, 15	D
4, 12	$\bar{A}\bar{B}C$

Similar solutions for *X*, *Y*, and *Z* are shown in Figs. 4-18 through 4-20, respectively. The circuit implementation is shown in Fig. 4-21. If $\bar{A}$, $\bar{B}$, $\bar{C}$, and $\bar{D}$ are not available, inverters can be placed on *A*, *B*, *C*, and *D*.

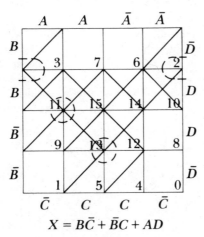

$$X = B\bar{C} + \bar{B}C + AD$$

Fig. 4-18   Karnaugh map of *X*.

CHECK THE SOLUTION: A table may be used to verify the findings of our Karnaugh map and circuit of Fig. 4-21. Again $\bar{W}$ will be used as an example, and the remainder of the variables can be solved for similarly. *W* is verified in Table 4-7 while variables *X*, *Y*, and *Z* are verified in Tables 4-8 through 4-10, respectively. Since $W = D + A\bar{C} + B\bar{C} + \bar{A}\bar{B}C$, we will find each of these separately and OR them.

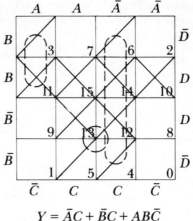

$$Y = \bar{A}C + \bar{B}C + AB\bar{C}$$

Fig. 4-19   Karnaugh map of Y.

$A\bar{C}$ is row 5 of the table and is found by ANDing A with $\bar{C}$. $B\bar{C}$ is found by ANDing B with $\bar{C}$ in row 6, and in row 7, $\overline{A}\bar{B}C$ is found by ANDing $\bar{A}$ with $\bar{B}$ with C. D is already row 4. The problem would be rather straightforward if we were not using NAND gates, but gates $G_1(A\bar{C})$, $G_2(B\bar{C})$, and $G_3(\overline{A}\bar{B}C)$ invert each of these variables. This is shown as rows 8, 9, 10, and 11 of Table 4-7 for $A\bar{C}$, $B\bar{C}$, $\overline{A}\bar{B}C$, and $\bar{D}$, respectively.

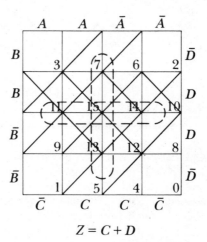

$$Z = C + D$$

Fig. 4-20   Karnaugh map of Z.

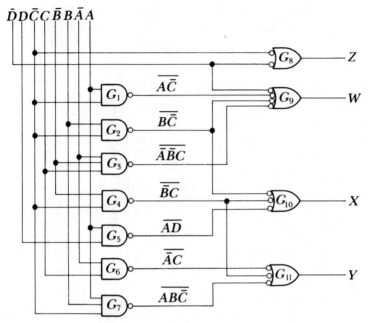

Fig. 4-21   Circuit implementation of W, X, Y, and Z.

TABLE 4-7   Table verification of W (Fig. 4-17)

A	0	1	0	1		0	1	0	1		0	1	0	1		0	1	0	1
B	0	0	1	1		0	0	1	1		0	0	1	1		0	0	1	1
C	0	0	0	0		1	1	1	1		0	0	0	0		1	1	1	1
D	0	0	0	0		0	0	0	0		1	1	1	1		1	1	1	1
$A\bar{C}$	0	1	0	1		0	0	0	0		0	1	0	1		0	0	0	0
$B\bar{C}$	0	0	1	1		0	0	0	0		0	0	1	1		0	0	0	0
$\bar{A}\bar{B}C$	0	0	0	0		1	0	0	0		0	0	0	0		1	0	0	0
$\overline{A\bar{C}}$	1	0	1	0		1	1	1	1		1	0	1	0		1	1	1	1
$\overline{B\bar{C}}$	1	1	0	0		1	1	1	1		1	1	0	0		1	1	1	1
$\overline{\bar{A}\bar{B}C}$	1	1	1	1		0	1	1	1		1	1	1	1		0	1	1	1
$\bar{D}$	1	1	1	1		1	1	1	1		0	0	0	0		0	0	0	0
$\bar{W}$	1	0	0	0		0	1	1	1		0	0	X	X		X	X	X	X
W	0	1	1	1		1	0	0	0		1	1	X	X		X	X	X	X

Now, $\overline{W}$ is found by ANDing these four functions as shown in row 12 and inverting to obtain W, shown as the last row. A similar procedure is used to check X, Y, and Z, as shown in Tables 4-8 through 4-10.

TABLE 4-8   Table verification of X (Fig. 4-18)

A	0	1	0	1	0	1	0	1	0	1	0	1	0	1	0	1
B	0	0	1	1	0	0	1	1	0	0	1	1	0	0	1	1
C	0	0	0	0	1	1	1	1	0	0	0	0	1	1	1	1
D	0	0	0	0	0	0	0	0	1	1	1	1	1	1	1	1
$B\bar{C}$	0	0	1	1	0	0	0	0	0	0	1	1	0	0	0	0
$\bar{B}C$	0	0	0	0	1	1	0	0	0	0	0	0	1	1	0	0
AD	0	0	0	0	0	0	0	0	0	1	0	1	0	1	0	1
$\overline{B\bar{C}}$	1	1	0	0	1	1	1	1	1	1	0	0	1	1	1	1
$\overline{\bar{B}C}$	1	1	1	1	0	0	1	1	1	1	1	1	0	0	1	1
$\overline{AD}$	1	1	1	1	1	1	1	1	1	0	1	0	1	0	1	0
$\bar{X}$	1	1	0	0	0	0	1	1	1	0	X	X	X	X	X	X
X	0	0	1	1	1	1	0	0	0	1	X	X	X	X	X	X

TABLE 4-9   Table verification of Y (Fig. 4-19)

A	0	1	0	1	0	1	0	1	0	1	0	1	0	1	0	1
B	0	0	1	1	0	0	1	1	0	0	1	1	0	0	1	1
C	0	0	0	0	1	1	1	1	0	0	0	0	1	1	1	1
D	0	0	0	0	0	0	0	0	1	1	1	1	1	1	1	1
$\bar{A}C$	0	0	0	0	1	0	1	0	0	0	0	0	1	0	1	0
$\bar{B}C$	0	0	0	0	1	1	0	0	0	0	0	0	1	1	0	0
$AB\bar{C}$	0	0	0	1	0	0	0	0	0	0	0	1	0	0	0	0
$\overline{\bar{A}C}$	1	1	1	1	0	1	0	1	1	1	1	1	0	1	0	1
$\overline{\bar{B}C}$	1	1	1	1	0	0	1	1	1	1	1	1	0	0	1	1
$\overline{AB\bar{C}}$	1	1	1	0	1	1	1	1	1	1	1	0	1	1	1	1
$\bar{Y}$	1	1	1	0	0	0	0	1	1	1	X	X	X	X	X	X
Y	0	0	0	1	1	1	1	0	0	0	X	X	X	X	X	X

TABLE 4-10   Table verification of Z (Fig. 4-20)

A	0 1 0 1	0 1 0 1	0 1 0 1	0 1 0 1
B	0 0 1 1	0 0 1 1	0 0 1 1	0 0 1 1
C	⌐0 0 0 0⌐	1 1 1 1	⌐0 0 0 0⌐	0 0 1 1
D	⟨⓪0 0 0	0 0 0 0⟩	1 1 1 1	1 1 1 1
$\bar{C}$	⌐1 1 1 1⌐	0 0 0 0	⌐1 1 1 1⌐	0 0 0 0
$\bar{D}$	⟨①1 1 1	1 1 1 1⟩	0 0 0 0	0 0 0 0
$\bar{Z}$	1 1 1 1	0 0 0 0	0 0 X X	X X X X
Z	0 0 0 0	1 1 1 1	1 1 X X	X X X X

PROBLEM 10

GIVEN: The Boolean expressions below.

$$F_1 = AB + ABC + BC$$
$$F_2 = (A\bar{B} + C)(A + \bar{B})C$$
$$F_3 = AB + (\bar{B} + \bar{C}) + \bar{A}C$$

FIND:

a.   Simplify the expression using Boolean algebra and implement using a maximum of two levels of logic.

b.   Use NAND gates to implement these functions and show the diagram.

SOLUTION:

a. $F_1 = AB + ABC + BC$

$F_1 = AB(1 + C) + BC = AB + BC = B(A + C)$

$F_2 = (A\bar{B} + C)(A + \bar{B})C = (A\bar{B}C + C)(A + \bar{B})$

$F_2 = C(A\bar{B} + 1)(A + \bar{B}) = C(A + \bar{B}) = AC + \bar{B}C$

$F_3 = AB + (\bar{B} + \bar{C}) + \bar{A}C = AB + A\bar{B} + \bar{A}\bar{C} + \bar{A}C$

$F_3 = A + \bar{A} = 1$

(*Note*: $\bar{B}$ implies $A\bar{B}$ and $\bar{C}$ implies $\bar{A}\bar{C}$.)

b.   The method for NAND gate implementation is to use AND gates and OR gates (no inversions) to implement the function and to place an inversion "ball" at the output of all AND gates and an inversion "ball" at the input of all AND gates used as OR functions.

NAND gates may be added where required to provide an inverter function. Figure 4-22 shows how F_1 is obtained using NAND logic, while Fig. 4-23 demonstrates the technique for F_2. F_3 is a trivial solution since it is a logic 1 level only.

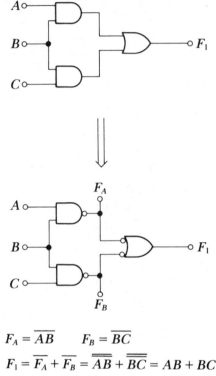

$$F_A = \overline{AB} \qquad F_B = \overline{BC}$$

$$F_1 = \overline{F_A} + \overline{F_B} = \overline{\overline{AB}} + \overline{\overline{BC}} = AB + BC$$

Fig. 4-22 Circuit for F_1 implementation.

CHECK THE SOLUTION: Use the Karnaugh map to check each function. The map for F_1 is shown in Fig. 4-23, and the simplified expression for F_1 can be determined by mapping AB by parallel horizontal lines, ABC by parallel vertical lines, and BC by parallel diagonal lines. From this figure, we verify that $F_1 = AB + BC$ since we select all the squares covered. The Karnaugh map for F_2 is shown in Fig. 4-25. $(A\overline{B} + C)$ is designated by horizontal parallel lines, $(A + \overline{B})$ by vertical parallel lines, and C by diagonal parallel lines. From this figure, we can determine F_2, by the common occupancy of squares by each ANDed function, to

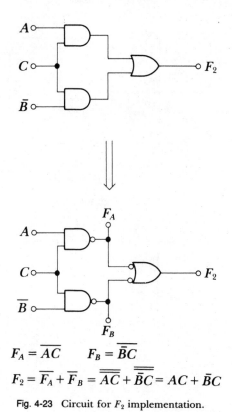

$$F_A = \overline{AC} \qquad F_B = \overline{\overline{B}C}$$

$$F_2 = \overline{F_A} + \overline{F_B} = \overline{\overline{AC}} + \overline{\overline{\overline{B}C}} = AC + \overline{B}C$$

Fig. 4-23   Circuit for F_2 implementation.

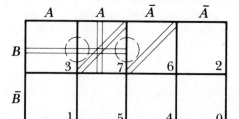

$$F_1 = AB + ABC + BC = AB + BC$$

Fig. 4-24   Karnaugh map for F_1.

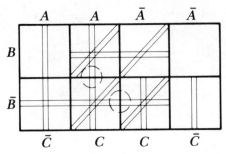

$$F_2 = (A\bar{B} + C)(A + \bar{B})C = AC + \bar{B}C$$

Fig. 4-25   Karnaugh map for F_2.

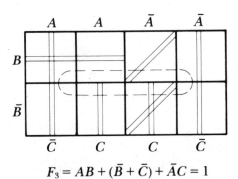

$$F_3 = AB + (\bar{B} + \bar{C}) + \bar{A}C = 1$$

Fig. 4-26   Karnaugh map for F_3.

be $AC + \bar{B}C$. The Karnaugh map for F_3 is shown in Fig. 4-26. AB is represented by horizontal parallel lines, $(\bar{B} + \bar{C})$ by vertical parallel lines, and $\bar{A}C$ by diagonal parallel lines. All squares being covered, our $F_3 = 1$ answer is verified.

PROBLEM 11

GIVEN: Four functions W, X, Y, Z that are functions of three inputs $A, B,$ and C as follows:

$$W = A + \bar{B}$$ $A = 0101 \quad 0101 \quad$ (LSB)

$$X = A\bar{B}$$ $B = 0011 \quad 0011$

$$Y = \bar{A}B\bar{C} + BC$$ $C = 0000 \quad 1111$

$$Z = A\bar{C}$$

FIND:

a. A function U that is a function of the four variables W, X, Y, Z, but we want U as a function of our original variables A, B, and C.

$$U = (WX + \bar{X}Y)\bar{Y}Z = f(A, B, C)$$

Implement using NAND logic.

b. If our output for U is any of the following, what could be wrong?

1. 1100 0000

2. 0100 0100

3. 1111 1111

SOLUTION:

a. The designation numbers and the evolution of the desired function U is shown in Table 4-11. We must find each variable W, X, Y, and Z separately, then combine them appropriately. Rows 1, 2, and 3 represent our three variables A, B, and C respectively. Row 4 represents W, which is found by ORing A with $\bar{B}$. $A + \bar{B}$ is

TABLE 4-11   Table for obtaining U as a function of A, B, and C.

A	0101	0101
B	0011	0011
C	0000	1111
$W = A + \bar{B}$	1101	1101
$X = A\bar{B}$	0100	0100
$Y = \bar{A}B\bar{C} + BC$	0010	0011
$Z = A\bar{C}$	0101	0000
WX	0100	0100
$\bar{X}Y$	0010	0011
$\bar{Y}Z$	0101	0000
$WX + \bar{X}Y$	0110	0111
$\bar{Y}Z$	0101	0000
$U = (WX + \bar{X}Y)\bar{Y}Z$	0100	0000
$U = A\bar{B}\bar{C}$		

represented by a designation number which has a 1 in either the A or $\bar{B}$ columns.

$$
\begin{array}{c|cccc|cccc}
A & 0 & 1 & 0 & 1 & 0 & 1 & 0 & 1 \\
\bar{B} & 1 & 1 & 0 & 0 & 1 & 1 & 0 & 0 \\
\hline
W = A + \bar{B} & 1 & 1 & 0 & 1 & 1 & 1 & 0 & 1
\end{array}
$$

Similarly for the X variable,

$$
\begin{array}{c|cccc|cccc}
A & 0 & 1 & 0 & 1 & 0 & 1 & 0 & 1 \\
\bar{B} & 1 & 1 & 0 & 0 & 1 & 1 & 0 & 0 \\
\hline
X = A\bar{B} & 0 & 1 & 0 & 0 & 0 & 1 & 0 & 0
\end{array}
$$

Since it is an AND function, we must have a 1 in both columns in order to produce a 1 in the X row. For the variable Y,

$$
\begin{array}{c|cccc|cccc}
\bar{A} & 1 & 0 & 1 & 0 & 1 & 0 & 1 & 0 \\
B & 0 & 0 & 1 & 1 & 0 & 0 & 1 & 1 \\
C & 0 & 0 & 0 & 0 & 1 & 1 & 1 & 1 \\
\bar{C} & 1 & 1 & 1 & 1 & 0 & 0 & 0 & 0 \\
\hline
\bar{A}B\bar{C} & 0 & 0 & 1 & 0 & 0 & 0 & 0 & 0 \\
BC & 0 & 0 & 0 & 0 & 0 & 0 & 1 & 1 \\
\hline
Y = \bar{A}B\bar{C} + BC & 0 & 0 & 1 & 0 & 0 & 0 & 1 & 1
\end{array}
$$

For the variable Z,

$$
\begin{array}{c|cccc|cccc}
A & 0 & 1 & 0 & 1 & 0 & 1 & 0 & 1 \\
\bar{C} & 1 & 1 & 1 & 1 & 0 & 0 & 0 & 0 \\
\hline
Z = A\bar{C} & 0 & 1 & 0 & 1 & 0 & 0 & 0 & 0
\end{array}
$$

Now we can determine our variable U as shown in Table 4-11 as rows 8 through 13. The circuit implementation of U is shown in Fig. 4-27.

b.

1. If we get the code 1100 0000, we can see from Table 4-11 that this corresponds to a code $U = \overline{BC}$, while we desire $U = ABC$. This says that A must be stuck at a 1 level.

2. A code of 0100 0100 corresponds to a function $U = A\bar{B}$, which indicates that C must be stuck at a 0 level.

3. If the output is always a 1 level, more than likely G_1 is stuck low or G_2 has failed and is stuck high. Another fault could be that the gates are operating properly and A, $\bar{B}$, and $\bar{C}$ are all stuck high, an unlikely occurrence.

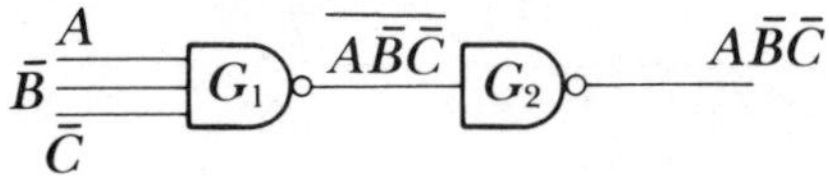

$$\begin{array}{c} A \\ B \\ C \end{array} \!\!-\!\!\boxed{G_1}\!\!-\!\!o\ \overline{A\overline{B}\overline{C}}\ \boxed{G_2}\!\!-\!\!o\qquad A\overline{B}\overline{C}$$

Fig. 4-27   Circuit implementation of U.

CHECK THE SOLUTION: Let's use a Karnaugh map to determine the designation number for each variable W and X, Y, and Z shown in Figs. 4-28 through 4-30, respectively. In Fig. 4-28, A is represented by parallel vertical lines and $\overline{B}$ by parallel horizontal lines. Since W is an OR function of $A + \overline{B}$, we want all squares occupied. The designation number can be found as follows: squares 2 and 6 are the only zeros; therefore our designation number is $W = 1101$ 1101. In Fig. 4-28 we can also determine our variable X, which is the common occupancy of A and $\overline{B}$. The designation number can be determined since only squares

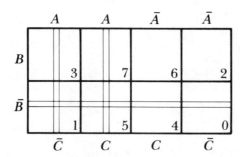

Fig. 4-28   Karnaugh map for determining W and X.

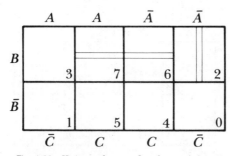

Fig. 4-29   Karnaugh map for determining Y.

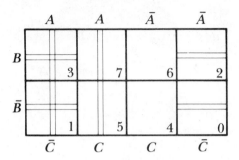

Fig. 4-30   Karnaugh map for determining Z.

1 and 5 are **1**. $X = 0100$ 0100. In Fig. 4-29, $\overline{A}B\overline{C}$ is represented by parallel vertical lines and BC by parallel horizontal lines. Squares 2, 6, and 7 are occupied (or function), so our designation number for Y is $Y = 0010$ 0011. In Fig. 4-30 we have the Karnaugh map for our variable Z. A is plotted as parallel vertical lines and $\overline{C}$ by parallel horizontal lines. Since Z is an AND function of A and $\overline{C}$, we want common occupancy as shown in squares 1 and 3. Therefore, our designation number for Z is 0101   0000. In Fig. 4-31 is a three-variable Karnaugh map for determining U. $(WX + \overline{X}Y)$ is plotted with parallel vertical lines while $\overline{Y}Z$ is plotted with parallel horizontal lines.

$$(WX + \overline{X}Y) = 0110 \quad 0111$$
$$\overline{Y}Z = 0101 \quad 0000$$

The common occupancy of these two functions is square 1, which is $A\overline{B}\overline{C}$, which checks our functions.

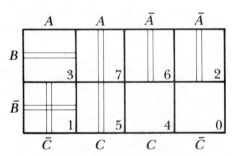

Fig. 4-31   Karnaugh map for determining U.

PROBLEM 12

GIVEN: Two types of integrated circuits are available for this design. One is a type D flip-flop with the truth table shown in Fig. 4-32, and the other is a NAND gate with its truth table shown in Fig. 4-33. There are two inputs, as shown in Fig. 4-32, with X being the data input and T being a synchronous clock input.

FIND: We desire the three waveforms shown in Fig. 4-34, one from each of the three D-type flip-flops. Indicate how you would do this using as few NAND gates and flip-flops as possible.

Q	0101	0101	Present state of Q
X	0011	0011	State of X
T	0000	1111	State of T (Clock)
Q'	0101	0011	Future state of Q

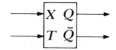

Fig. 4-32   Type D flip-flop.

SOLUTION: We can see that the waveforms are periodic with a period equal to five clock pulses. We can assign a designation number for each of the three waveforms desired, as illustrated in Fig. 4-35.

We can assign a state diagram to this problem making A the least significant bit (LSB). Column 1 corresponds to a logic count of 1 while columns 2 through 5 correspond to counts of 7, 3, 6, and 4, respectively. Our state diagram is shown in Fig. 4-36.

A	01	01	Input A
B	00	11	Input B
	11	10	Output Y

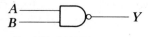

Fig. 4-33   NAND gate.

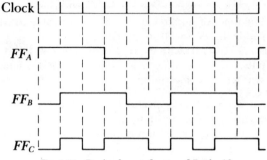

Fig. 4-34   Desired waveforms of Prob. 12.

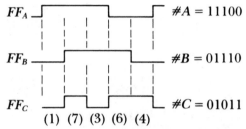

$\#A = 11100$

$\#B = 01110$

$\#C = 01011$

(1) (7) (3) (6) (4)

Fig. 4-35   Designation number assignments of Prob. 12.

Let's generate our truth table (standard basis) which shows the present and future states of the counter. Designate constrained states (unwanted) with X's in the designation number.

Present state	A	0 1 0 1	0 1 0 1
	B	0 0 1 1	0 0 1 1
	C	0 0 0 0	1 1 1 1
		0 1 2 3	4 5 6 7
		↓↓↓↓	↓↓↓↓
		$X7X6$	$1X43$
Future state	A'	$X1X0$	$1X01$
	B'	$X1X1$	$0X01$
	C'	$X1X1$	$0X10$

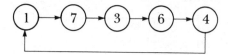

Fig. 4-36   State diagram for Prob. 12.

We know from the truth table for the D flip-flop (Fig. 4-32) that whatever is on the X input is transferred to the Q output whenever we get a clock pulse (T). Consequently, if we look at the previous states of the flip-flops, and gate them in the proper sequence into the T input, we can get the desired response. We can also make use of redundancy to minimize gate functions. Let's plot each future state on a Karnaugh map to get an expression for A', B', and C'. On Figs. 4-37 through 4-39 we map the designation number that yields the minimized Boolean expression shown on the figures representing our desired variables. A slanted line represents a **1** while X's represent constrained conditions

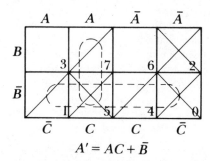

$$A' = AC + \bar{B}$$

Fig. 4-37   Karnaugh map of A'.

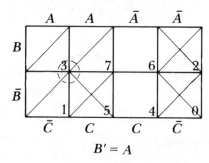

$$B' = A$$

Fig. 4-38   Karnaugh map of B'.

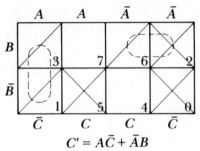

$$C' = A\bar{C} + \bar{A}B$$

Fig. 4-39   Karnaugh map of C'.

which can be taken as a **1** or **0** to enhance minimization. The numbers in each square represent the binary equivalent of A, B, and C. (*Example*: $\bar{A}B\bar{C} = 2$.)

Now that we have minimized the functions A', B', and C', we can use these as inputs to the X input of each flip-flop and obtain our desired counting sequence. The minimization is shown on each figure. See Fig. 4-40 for the overall schematic. The best way to easily remember the gating is to realize that a **0** on any input of a NAND gate will give a **1** out, and a **0** output is only realized when all inputs are **1** levels. The defining expression for a NAND gate with inputs A and B is $f = \overline{A \cdot B} = \bar{A} + \bar{B}$. All gates shown are NAND gates. When we activate a NAND gate implemented in an AND function, we will assume its output is **0**. When we activate a NAND gate implemented in an OR function we will assume its output is a **1**.

CHECK THE SOLUTION: First let's see what happens when we come up in one of our constrained states: 0, 2, or 5. Our new designation numbers for A', B', and C' can be determined by inspection:

A	0101	0101
B	0011	0011
C	0000	1111
$A' = AC + \bar{B}$	1100	1101
$B' = A + \bar{C}$	0101	0101
$C' = A\bar{C} + \bar{A}B$	0111	0010

Thus we can determine if we come up in state **0** for $A/B/C$, that $A'/B'/C'$ becomes a 1; for $A/B/C = 2$, $A'/B'/C'$ will come up to 4; and

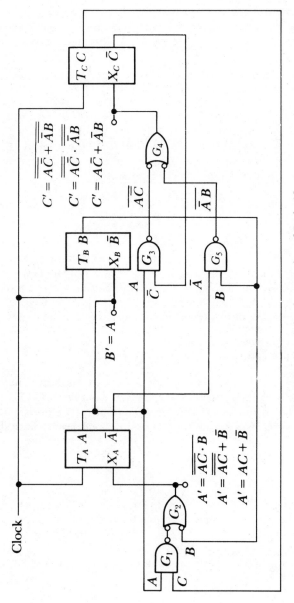

Fig. 4-40 Circuit implementation of Prob. 12.

213

for $A/B/C = 5$, $A'/B'/C'$ becomes a 1. These are all proper numbers in the counting sequence, and we will not reach a lockout condition.

To check our schematic solution, let's go through each count sequence to ensure that we have gated the X inputs correctly:

a. Present state of flip-flops: 1   0   0   ①
    Future state of flip-flops: 1   1   1   ⑦

If we have a count of 1 in the counter, $A = 1$, $B = C = 0$. Since $X_B = A$, $X_B = 1$. Since B is a **0**, the output of $G_2(X_A)$ is a **1**. Since $A = \overline{C} = 1$, the output of G_3 is a **0** and the output of $G_4(X_C)$ is a **1**. Therefore, since $X_A = X_B = X_C = 1$, we will have a count of 7 in the counter at the next clock pulse.

b. Present state of flip-flops: 1   1   1   ⑦
    Future state of flip-flops: 1   1   0   ③

$A = B = C = 1$; therefore, X_B is a **1** since A is a **1**. Since A and C are **1**, the output of G_1 is a **0**, causing the output of $G_2(X_A)$ to be a **1**. $\overline{A}$ is a **0**, causing the output of G_5 to be a **1**. The output of G_3 is a **1** since one input $\overline{C}$ is a **0**. Since both inputs to G_4 are **1** levels, the output of $G_4(X_C)$ is a **0**. Since $X_A = X_B = 1$, and $X_C = 0$, we will have a count of 3 in our counter after the next clock arrives.

c. Present state of flip-flops: 1   1   0   ③
    Future state of flip-flops: 0   1   1   ⑥

$A = B = 1$, and $C = 0$. The output of G_1 is a **1** since the C input is a **0**. Since both inputs to G_2 are **1** levels, the output of $G_2(X_A)$ is a **0**. Since the $\overline{A}$ input to G_5 is a **0**, the output of $G_5(X_B)$ is a **1**. The output of G_6 is a **0** since both A and $\overline{C}$ are **1** levels. Since the G_6

d. Present state of flip-flops: 0   1   1   ⑥
    Future state of flip-flops: 0   0   1   ④

$A = 0$, $B = C = 1$. $X_B = A = 0$. Since A is a **0**, the output of G_1 is a **1**. Since the other input (B) to G_2 is also a **1**, the output of $G_2(X_A)$ is a **0**. Because both $\overline{A}$ and B are **1** levels, the output of G_5 is a **0** causing the output of $G_4(X_C)$ to be a **1**. Thus, $X_A = X_B = 0$ and $X_C = 1$, which will cause the counter to assume a count of 4 on the next clock pulse.

e. Present state of flip-flops: 0   0   1   ④
    Future state of flip-flops: 1   0   0   ①

$A = B = 0$ and $C = 1$. $X_B = A = 0$. Since $B = 0$, the output of $G_2(X_a)$ is a **1**. The output of G_3 is **1** since $A = 0$, and the output of G_5 is a **1** since $B = 0$. Now, since both inputs to G_4 are **1** levels, the output of G_4 is a **0**. Since $X_A = 1$, and $X_B = X_C = 0$, we will have a count of 1 in the counter at the next clock pulse. Our solution check is therefore complete.

PROBLEM 13 _____

GIVEN: The input functions A, B, C, and D shown below:

A 0101   0101   0101   0101

B 0011   0011   0011   0011

C 0000   1111   0000   1111

D 0000   0000   1111   1111

FIND: Implement, using the minimum number of input + output combinations possible for gating, the following functions of A, B, C, and D:

W 0101   0011   1100   0111

X 1101   0011   1100   0111

Y 0101   0001   1100   0111

Z 1010   1110   0111   1000

SOLUTION: We might solve by creating our Karnaugh map and then minimizing using conventional techniques. However, closer attention to the problem reveals that $X = f(W) = f(Y) = f(A, B, C, D)$, and we can use pyramiding techniques to simplify the solution. Another simplification can be done by inverting Z. $\bar{Z}$ then becomes $\bar{Z} =$ 0101 0001 1000 0111 and $Y = f(Z)$. Therefore we can use the following pyramiding:

$$\bar{Z} \longrightarrow Y \longrightarrow W \longrightarrow X$$

We can simplify by making $\bar{Z}$ assume constrained states on our Karnaugh map (X's) and plotting Y over it and minimizing. This is done in Fig. 4-41. We obtain Y as a function of $\bar{Z}$: $Y = \bar{Z} + A\bar{B}D$. Now, in

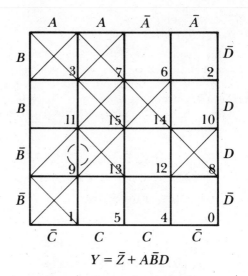

$$Y = \bar{Z} + A\bar{B}D$$

Fig. 4-41 Karnaugh map for Y.

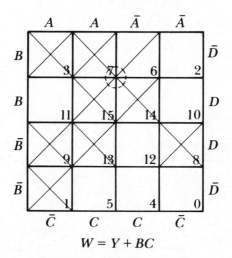

$$W = Y + BC$$

Fig. 4-42 Karnaugh map for W.

216

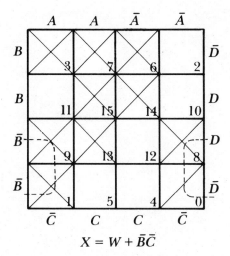

$$X = W + \bar{B}\bar{C}$$

Fig. 4-43 Karnaugh map for X.

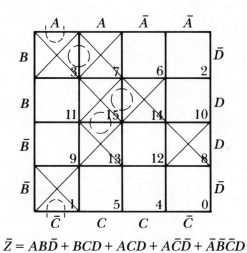

$$\bar{Z} = AB\bar{D} + BCD + ACD + A\bar{C}\bar{D} + \bar{A}\bar{B}\bar{C}D$$

Fig. 4-44 Karnaugh map for $\bar{Z}$.

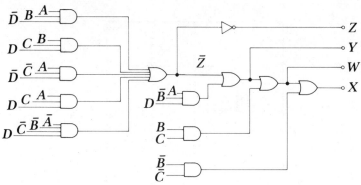

Fig. 4-45   Circuit implementation for Prob. 13.

Fig. 4-42, we can plot Y as constrained states and plot W in the remaining unconstrained boxes. We obtain $W = Y + BC$. Similarly, we can plot W in Fig. 4-43 and obtain $X = W + \overline{BC}$. Now all that remains is to simplify $\overline{Z}$. The Karnaugh map for $\overline{Z}$ is shown in Fig. 4-44, and its minterm representation is $\overline{Z} = AB\overline{D} + BCD + ACD + \overline{A}\,\overline{C}\,\overline{D} + \overline{A}\,\overline{B}\,\overline{C}\,D$. The circuit implementation is shown in Fig. 4-45.

CHECK THE SOLUTION: The solution can best be checked by checking each step. No other check seems readily available for this particular solution.

PROBLEM 14 _____

GIVEN: It is desired to take a continuous clock input of 1 kHz and to divide its period into 45° increments. Explain how you would do this and provide a block diagram showing your approach. The outputs are to be synchronous with the input.

SOLUTION: A convenient approach to this problem is to use a phase-lock-loop. A conventional phase-lock-loop adjusts its frequency to make the feedback waveform (ϕ_ε) 90° out of phase with the input and identical in frequency. The 45° waveforms desired are shown as A through H, respectively, in Fig. 4-46. The block diagram for this approach is shown in Fig. 4-47.

The operation of the loop is as follows. Since feedback is used, the loop adjusts to make the waveforms of the clock and ϕ_ε be 90° out of phase. The low-pass filter takes the output from the phase comparator and filters out the dc component that in turn controls the frequency of

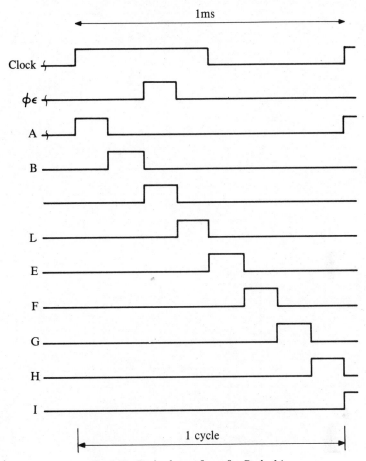

Fig. 4-46   Desired waveform for Prob. 14.

the VCO (voltage controlled oscillator). The output frequency of the oscillator (CL) runs at 8 kHz to provide the eight 45° increments. This technique is also convenient for multiplying input signals synchronously.

PROBLEM 15

GIVEN: The parallel A/D converter shown in Fig. 4-48.

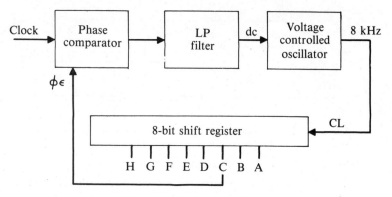

Fig. 4-47 Solution to Prob. 14.

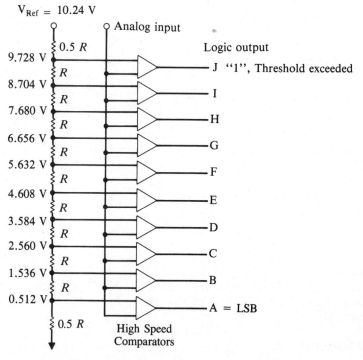

Fig. 4-48 Parallel A/D converter for Prob. 15.

FIND:

a. Why would you use this type of A/D converter over other types such as successive approximation or ramp?

b. What is the quantizing error for this converter?

c. Show the gating scheme to make the outputs binary.

SOLUTION:

a. A parallel A/D converter is used because of its real-time speed which can be in the nanosecond range for some of the state-of-the-art converters. Four-bit converters are easily made in the 100-ns conversion range.

b. Quantizing error is that error introduced by the finite resolution determined by the number of bits selected. If Fig. 4-48 were a one-bit converter, half scale would be 5.12 V. At 5.13 V we would have a HI output and at 5.11 V we would have a LO output. Therefore, the error is one-half the bit; or more generally, one-half the LSB. The quantizing error for this problem is 0.512 V out of 10.24, or 5%.

c. The function diagram for the input/output relationship is shown in Table 4-12. W, X, Y, Z represent the binary outputs; W is the LSB. From inspection of the A through H outputs of the converter, the binary representation can be determined. The boxed states are the only time that these states occur from the comparators. The equa-

TABLE 4-12 Comparator outputs and binary representation for Prob. 15

A	0	1	1	1	1	1	1	1	1	1	1	LSB	
B	0	0	1	1	1	1	1	1	1	1	1		
C	0	0	0	1	1	1	1	1	1	1	1		
D	0	0	0	0	1	1	1	1	1	1	1		
E	0	0	0	0	0	1	1	1	1	1	1		
F	0	0	0	0	0	0	1	1	1	1	1	A/D outputs	
G	0	0	0	0	0	0	0	1	1	1	1		
H	0	0	0	0	0	0	0	0	1	1	1		
I	0	0	0	0	0	0	0	0	0	1	1		
J	0	0	0	0	0	0	0	0	0	0	1	MSB	
W	0	1	0	1	0	1	0	1	0	1	0	LSB	
X	0	0	1	1	0	0	1	1	0	0	1	Binary representation	
Y	0	0	0	0	1	1	1	1	0	0	0		
Z	0	0	0	0	0	0	0	0	1	1	1	MSB	

tions for the solutions are derived and shown in Table 4-13, and the gating structure is shown in Fig. 4-49. DeMorgan's theorem is used to find the final solution (i.e., $\overline{AB \cdots Z} = \overline{A} + \overline{B} + \cdots + \overline{Z}$). Karnaugh maps can also be used to find the gating.

TABLE 4-13   Output equations for Prob. 15

$$W = A\overline{B} + C\overline{D} + E\overline{F} + G\overline{H} + I\overline{J}$$
$$X = B\overline{C} + C\overline{D} + F\overline{G} + G\overline{H} + IJ$$
$$Y = D\overline{E} + E\overline{F} + F\overline{G} + G\overline{H}$$
$$Z = H\overline{I} + I\overline{J} + IJ = H\overline{I} + I(\overline{J} + J) = H\overline{I} + I$$

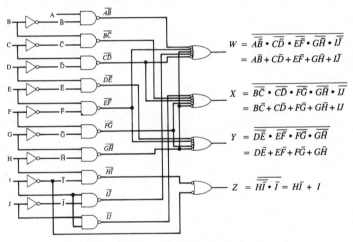

Fig. 4-49   Circuit solution to Prob. 15.

PROBLEM 16

GIVEN: The circuit of Fig. 4-50 with the significant-error contributors determined to be R_1, R_2, R_3, R_4, and V_Z.

FIND: Determine the statistical maximum output tolerance and the worst-case maximum output tolerance and explain the difference. What advantages can be realized using a statistical approach? What conclusions can be drawn if the distributions of the component values are not Gaussian? A root-sum-of-squares (RSS) statistical approach is recommended.

SOLUTION: The equation that relates the output to the input of the basic op-amp circuit in Fig. 4-50 is provided in Appendix C3.

$$E_o = \frac{(V_1 - V_2)(R_3 + R_4)R_2}{(R_1 + R_2)R_3} + \frac{(R_3 + R_4)V_2}{R_3} - \frac{V_3 R_4}{R_3} \quad \text{(C3-1)}$$

For this problem $V_1 = V_Z$, $V_2 = V_3 = 0$. Therefore:

$$E_o^* = \frac{R_2(R_3 + R_4)V_Z}{R_3(R_1 + R_2)} = \frac{2.2 \text{ k}\Omega(19.7 \text{ k}\Omega)5 \text{ V}}{4.7 \text{ k}\Omega(12.2 \text{ k}\Omega)} = 3.779 \text{ V dc}$$

*means nominal; " means maximum; ' means minimum.

Now let's determine the statistical maximum output. It can be determined by inspection or by calculation that the output increases with increasing values of V_Z, R_2, and R_4; and decreasing value of R_1 and R_3. Calculate the value of E_o for each variable with its allowable tolerance.

$$E_{o1}'' = \frac{(2.2 \text{ k}\Omega)(19.7 \text{ k}\Omega)5 \text{ V}}{4.7 \text{ k}\Omega(9 \text{ k}\Omega + 2.2 \text{ k}\Omega)} = 4.117 \text{ V dc} \quad \text{(due to } R_1)$$

$$\Delta E_{o1}'' = E_{o1}'' - E_o^* = 4.117 - 3.779 = 0.388 \text{ V dc}$$

$$E_{o2}'' = \frac{(2.4 \text{ k}\Omega)(19.7 \text{ k}\Omega)5 \text{ V}}{4.7 \text{ k}\Omega(10 \text{ k}\Omega + 2.4 \text{ k}\Omega)} = 4.056 \text{ V dc} \quad \text{(due to } R_2)$$

$$\Delta E_{o2}'' = E_{o2}'' - E_o^* = 0.227 \text{ V dc}$$

$$E_{o3}'' = \frac{(2.2 \text{ k}\Omega)(4.23 \text{ k}\Omega + 15 \text{ k}\Omega)5 \text{ V}}{4.23 \text{ k}\Omega(12.2 \text{ k}\Omega)} = 4.099 \text{ V dc (due to } R_3)$$

$$\Delta E_{o3}'' = E_{o3}'' - E_o^* = 0.320 \text{ V dc}$$

$$E_{o4}'' = \frac{(2.2 \text{ k}\Omega)(4.7 \text{ k}\Omega + 16 \text{ k}\Omega)5 \text{ V}}{4.7 \text{ k}\Omega(12.2 \text{ k}\Omega)} = 3.971 \text{ V dc} \quad \text{(due to } R_4)$$

$$\Delta E_{o4}'' = 0.192 \text{ V dc}$$

$$E_{o5}'' = E_o^* \frac{(5.05)}{5} = 3.817 \text{ V dc} \quad \text{(due to } V_Z)$$

$$\Delta E_{o5}'' = E_{o5}'' - E_o^* = 0.038 \text{ V dc}$$

$$\Delta E_{o|RSS} = [(\Delta E_{o1}'')^2 + (\Delta E_{o2}'')^2 + (\Delta E_{o3}'')^2 + (\Delta E_{o4}'')^2 + (\Delta E_{o5}'')^2]^{1/2}$$
$$= 0.585 \text{ V dc}$$

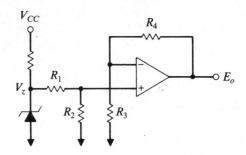

$$V_z = 5.0 \pm 0.05 \text{ V} \qquad R_3 = 4.7 \pm 0.47 \text{ k}\Omega$$

$$R_1 = 10 \pm 1 \text{ k}\Omega \qquad R_4 = 15 \pm 1 \text{ k}\Omega$$

$$R_2 = 2.2 \pm 0.2 \text{ k}\Omega$$

Fig. 4-50 Circuit configurations for Prob. 16.

Our worst-case maximum is determined by allowing each variable to assume its worst-case value.

$$E_{o(\text{max})} = \frac{(2.4 \text{ k}\Omega)(20.23 \text{ k}\Omega)(5.05 \text{ V})}{4.23 \text{ k}\Omega(11.4 \text{ k}\Omega)} = 5.085 \text{ V dc}$$

and

$$\Delta E_{o|\text{worstcase}} = 5.085 - 3.779 = 1.306 \text{ V dc}$$

which is a very unrealistic number to have to live with. The RSS analysis is more realistic and is based upon the assumption that all components have Gaussian normal distributions about their means as shown in Fig. 4-51. When we find the RSS value, we are merely finding the equivalent 3σ (3 standard deviations) point for multiple variables. If we do not have Gaussian distributions for each of our variables, the analysis may not be absolutely correct for small samples. However, there is the central limit theorem, which states that as the number of variables becomes large, the total distribution approaches Gaussian no matter what the shape of their individual distributions. Since it is almost certain that many of our components are not Gaussian, the final value theorem tells us why our analysis may have been successful in many practical designs. Why might our distributions not be Gaussian? As an example, $\pm 1\%$ values may be selected from a particular batch which omits the center section out of the parts shipped as $\pm 5\%$. The mean remains the same while the 3σ point moves away from the mean.

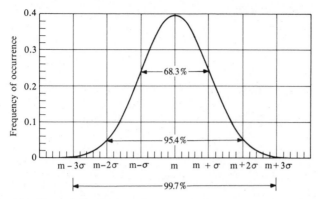

Fig. 4-51   Normal distribution of random values around the mean value.

The use of statistical analyses to predict performance is a preferred approach because it can reduce cost in manufacture and test (yield) and may reduce size, weight, and power while increasing reliability. The test issue is one of significant proportion because a unit under test approaching worst-case limits must surely have a failure associated with it. The rule is to test as tightly as possible without suffering significant loss of yield, and RSS limits offer the optimum choice for this condition.

PROBLEM 17

GIVEN: A sinusoidal oscillator is to be designed with the following characteristics:

Frequency:	1800 +/− 0.01%
Amplitude:	20 V_{p-p} +/− 0.1%
Load:	600 Ω
Temperature:	−55°C to 100°C
Harmonic distortion:	0.1% or less

FIND: Since the specifications are rather stringent, explain in block form how you would design the oscillator. Space is limited and does not permit the use of crystals resonant at frequencies less than 1 MHz.

SOLUTION: The block diagram for a possible solution is shown in Fig. 4-52. The accuracy of the oscillator frequency rules out the use of capacitors and inductors for the frequency-determining elements due to their loose tolerances. (Precision capacitors and inductors are large

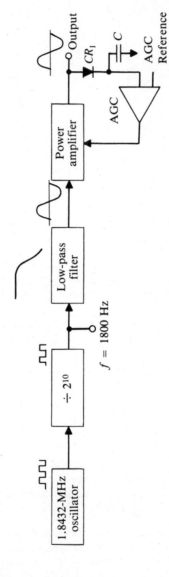

Fig. 4-52 Block diagram of 1800-Hz sinusoidal oscillator.

and costly and usually cannot be guaranteed much better than $+/-$ 1%, even over a limited temperature range.) Yet we are restricted from using an 1800-Hz crystal. Instead, we'll have to use a high-frequency crystal and count down using digital counters.

The problem with this solution is that we need a sharp roll-off filter that passes the fundamental at 1800 Hz (a pure sinusoid) yet provides greater than 60 dB (1000:1) attenuation at 5400 Hz. The filter must be constructed with high-precision capacitors which are large and costly, but size was given as a constraint. The high precision is required to keep the gain variation down, in order to make the AGC design realistic and achievable.

A better solution to this problem is shown in Fig. 4-53. Here we use a ROM that stores the sine function from 0 to 90° in binary form and is followed by a D/A converter that converts the binary information into a sine wave. (The steps are not visible on an oscilloscope if the number of bits are large.) The clock is followed by an up/down counter that counts up from zero to a count of 1023 (90°), and back down to a count of zero (180°). Then the sign input on the D/A converter is activated by the flip-flop and the up/down cycle is completed again to provide for the 180° to 360° negative portion.

If we assume that we have a triangular wave riding on the sine wave, the worst-case rms distortion is 0.577 times the peak value of the triangular wave. Refer to Fig. 4-54. Let E_B be the magnitude of the LSB. The peak value of the distortion is $E_P = E_B/2(\cos 45°) = 0.35\,E_B$; and the rms value is $0.577\,E_P = 0.175\,E_B$, or approximately 0.2 of the LSB magnitude. One LSB $\approx 0.1\%$; therefore, our distortion is less than 0.02%.

(Note the distortion should be significantly less because the $\Delta V/\Delta t$ gets much smaller as we approach the peak of the sine wave. The rms value of a triangular wave is 0.577 its peak value.)

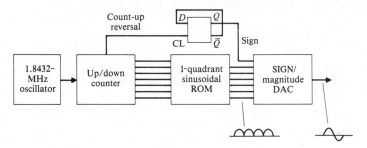

Fig. 4-53   Solution to Prob. 17.

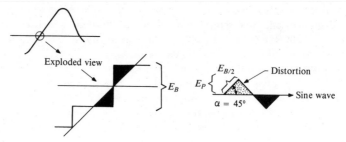

Fig. 4-54   Determining distortion for Prob. 17.

PROBLEM 18

GIVEN: The circuit diagram of a communication link as shown in Fig. 4-55. The input to the amplifier is 26 dBm (referenced to 1 mW) and its gain is 12 dB. The output of the amplifier is connected to a telephone line which has an efficiency of 80% and is terminated with a transformer whose loss is 1.5 W. The secondary load on the transformer is 10 Ω.

FIND: What is the voltage appearing across the load?

SOLUTION: Find the input power to the amplifier and label this P_1:

$$26\ \text{dBm} = 10 \log P_1/0.001$$
$$P_1/0.001 = 10^{26/10} = 398.1$$
$$P_1 = 0.3981\ \text{W}$$

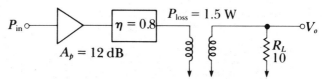

Fig. 4-55   Block diagram of communication link.

Find the power out of the amplifier and label it P_2:

$$12\ \text{dB} = 10 \log P_2/P_1$$
$$P_2/P_1 = 10^{12/10} = 15.85$$
$$P_2 = 6.31\ \text{W}$$

Label the input to the transformer P_3 and calculate the power at this point:

$$P_3 = \eta P_2 \quad \eta = \text{efficiency}$$
$$= (0.8)(6.31) = 5.05 \text{ W}$$

Designate the power delivered to the load P_4, and calculate:

$$P_4 = P_3 - 1.5 \text{ W}$$
$$P_4 = 3.55 \text{ W}$$

Now we can find the voltage across the load resistor as follows:

$$V_o^2/R_L = 3.55 \text{ W}$$
$$V_o = 5.96 \text{ V rms}$$

CHECK THE SOLUTION: Check each step numerically.

PROBLEM 19

GIVEN: A communication transmission line with the following characteristics:

$$R = 100 \ \Omega/\text{mi}$$
$$L = 1 \text{ mH/mi}$$
$$C = 0.1 \ \mu\text{F/mi}$$
$$G = 0.1 \ \mu\text{mho/mi}$$

FIND: If the line is terminated in its characteristic impedance and the frequency is 10 kHz, calculate:

a.   Characteristic impedance Z_o in ohms

b.   Propagation constant γ

c.   Attenuation α in dB

d.   Phase shift β in degrees

e.   Velocity v

f.   Wavelength λ

SOLUTION: To find Z_o and λ we must first find Z, which is the total series impedance of the line, and Y, which is the total shunt admittance of the line.

$$Z = R + j\omega L = 100 + j(2\pi)(10^4)(10^{-3})$$
$$= 100 + j62.8 = 118.1\underline{/32.1°}$$
$$Y = G + jwC = 10^{-7} + j(2\pi)(10^4)10^{-7} = j0.0063$$
$$= 6.3 \times 10^{-3}\underline{/90°}$$

The characteristic impedance of the line can now be calculated:

$$Z_o = \left(\frac{Z}{Y}\right)^{1/2} = \left(\frac{118.1\underline{/32.1^\circ}}{6.3 \times 10^{-3}\underline{/90^\circ}}\right)^{1/2} = 136.9\underline{/-29^\circ}$$

The propagation constant can be found as follows:

$$\gamma = (ZY)^{1/2} = [(118.1\underline{/32.1^\circ})(6.3 \times 10^{-3}\underline{/90^\circ})]^{1/2} = 0.863\underline{/61.1^\circ}$$

$$= 0.417 + j.756$$

But $\gamma = \alpha + j\beta$; therefore

$$\alpha = 0.417 \text{ neper/mi}$$

$$\beta = 0.756 \text{ rad/mi}$$

We desire α in decibels (dB); we can convert as follows:

$$\alpha = 0.417 \text{ neper/mi} \times 8.686 \text{ dB/neper} = 3.62 \text{ dB}$$

The velocity can now be determined:

$$v = \omega/\beta = (2\pi)(10^4)/0.756 = 8.31 \times 10^4 \text{ mi/s}$$

$$\lambda = (2\pi)/\beta = (2\pi)/0.756 = 8.31 \text{ mi}$$

CHECK THE SOLUTION: Check each step for numerical accuracy.

PROBLEM 20

GIVEN: A communication link between a satellite and a ground station at a separation distance of $d = 30,000$ km.

FIND:

a.   The maximum power density reaching the ground station if the directive transmitter power is $P_T = 15$ W and has an antenna of cross-sectional area $A_T = 1.15$ m^2. Assume isotropic transmission. The frequency of transmission is 500 MHz, and the atmospheric attenuation is determined to be 50 dB.

b.   The power received by the ground receiver if its antenna has a directive gain $G_R = 30$ dB and its area is 300 m^2.

c.   If the bandwidth of the receiver is 10 kHz and the space noise effective temperature is $T_e = 1000$ K, what is the background noise power received by the ground antenna?

d.   The signal-to-noise ratio in decibels.

SOLUTION:

a.   The model for our system can be described by the block diagram of Fig. 4-56. We can develop the equations for each power, or power density, with magnitudes referenced with respect to the receiver. P_1 represents the power density at the receiver as a result of isotropic radiation, with the ground site a part of a sphere of radius d.

$$P_1 = \frac{P_T}{4\pi d^2} = G_1 P_T \text{ W/m}^2 \qquad G_1 = \frac{1}{4\pi d^2}$$

We have a directional gain of the transmitter represented by

$$P_2 = G_2 P_1 = \frac{4\pi A_T P_1}{\lambda^2} \text{ W/m}^2 \qquad G_2 = \frac{4\pi A_T}{\lambda^2}$$

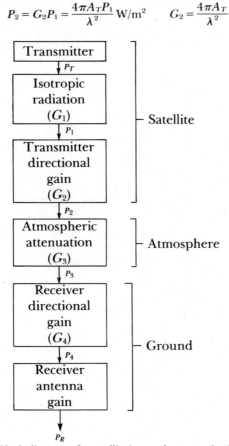

Fig. 4-56   Block diagram of a satellite/ground communication link.

Atmospheric attenuation can be included as follows:

$$G_3 = -50 \text{ dB} = 10 \log \left(\frac{P_3}{P_2}\right)$$

$$-5 = \log \left(\frac{P_3}{P_2}\right)$$

$$\frac{P_3}{P_2} = 10^{-5}$$

$$G_3 = 10^{-5}$$

Similarly, the receiver directional gain can be calculated:

$$G_4 = 30 \text{ dB} = 10 \log \left(\frac{P_4}{P_3}\right)$$

$$\frac{P_4}{P_3} = 10^3 = G_4$$

$$P_4 = G_4 P_3$$

The power P_5 reaching the receiver is nothing more than the receiver area times the power density P_4:

$$P_R = A_R P_4 \text{ W}$$

The power density reaching the ground excludes the receiver directional gain and receiver antenna gain and can be calculated as follows (*note*: $\lambda = (3 \times 10^8)/f = (3 \times 10^8)/(5 \times 10^8) = 0.6$ m):

$$P_3 = G_3 P_2 = G_3 G_2 P_1 = G_1 G_2 G_3 P_T$$

$$= \left(\frac{1}{4\pi d^2}\right)\left(\frac{4\pi A_T}{\lambda^2}\right)(10^{-5})P_T$$

$$= \frac{10^{-5} A_T P_T}{\lambda^2 d^2} \text{ W/m}^2$$

$$= \frac{(10^{-5})(1.15 \text{ m}^2)(15 \text{ W})}{(0.6 \text{ m})^2 (3 \times 10^7 \text{ m})^2} = 5.32 \times 10^{-19} \text{ W/m}^2$$

b. The power at the input to the receiver is

$$P_R = A_R P_4 = A_R G_4 P_3$$
$$= (300 \text{ m}^2)(10^3)(5.32 \times 10^{-19})$$
$$= 1.6 \times 10^{-13} \text{ W}$$

c.   The expression for the background noise power is:

$P_N = KT_eB$

K = Boltzmann's constant; 1.38×10^{-23} J/K

B = noise bandwidth; assume the bandwidth given is the noise bandwidth, since no other information is given.

T_e = effective noise temperature

$P_N = (1.38 \times 10^{-23})(10^3)(10^4) = 1.38 \times 10^{-16}$ W

d. $\dfrac{S}{N} = \dfrac{P_R}{P_N} = \dfrac{1.6 \times 10^{-13}}{1.38 \times 10^{-16}} = 1159.4$ W/W

$\dfrac{S}{N} = 10 \log (1159.4) = 30.64$ dB

CHECK THE SOLUTION: Check each step numerically.

PROBLEM 21 _____

GIVEN: Assume that we are given an analog message $m(t)$ that is band-limited with a highest available frequency of f_M lasting for T seconds.

FIND:

a.   If we sample and convert to a digital format using an A/D converter with N quantizing levels, find the channel capacity required to transmit the signal with no loss of information.

b.   The channel capacity required for a pulse-code modulation (PCM) coding technique with M possible amplitude values for each bit and m code elements. Calculate the bandwidth if the transmission is binary with five bits of accuracy and $f_M = 2$ kHz.

c.   If we have noise in the system and a signal-to-noise ratio of $S/N = 5$, determine the channel capacity for a system having a bandwidth of 5 kHz.

SOLUTION:

a.   The rate at which we sample without loss of information is called the *Nyquist rate* and is twice the maximum signal rate, or a frequency f_s of

$$f_s = 2f_M$$

Also, if we use an A/D converter with N possible quantizing levels, we need $\log_2 N$ bits for binary encoding of every sample. This imposes the following transmission rate on our system:

$$C = 2f_M \log_2 N \text{ bit/s}$$

b. The number of quantizing levels for this condition is determined as follows:

$$N = M^m \text{ possible quantizing levels}$$

Substituting $N = M^m$ into our previous equation for C, we find

$$C = 2f_M m \log_2 M \text{ bit/s}$$

This assumes that the quantizer has a symmetrical capability about zero, so that the M levels will be distributed equally above and below zero. For the given conditions ($m = 5$, $M = 2$),

$$C = 2(2 \times 10^3)5 \log_2 2$$
$$= 20 \text{ kHz}$$

c. Using the Hartley-Shannon law, we can determine the channel capacity when noise is present as follows:

$$C = W \log_2 (1 + S/N)$$
$$= 5 \times 10^3 \log_2 6 = (5 \times 10^3)(2.585) = 12.92 \text{ kHz}$$

Note:

$$\log_2 y = x$$
$$y = 2^x$$
$$\ln y = \ln 2^x = x \ln 2$$
$$x = \frac{\ln y}{\ln 2}$$

For this problem,

$$x = \frac{\ln 6}{\ln 2} = \frac{1.79}{0.693} = 2.582$$

CHECK THE SOLUTION: Check each step for numerical accuracy.

PROBLEM 22

GIVEN: A microwave transmitter shown in Fig. 4-57 that operates in an assigned band from 5.5 to 6 GHz. Each component block possesses

$f_m = 0$ to 5 MHz

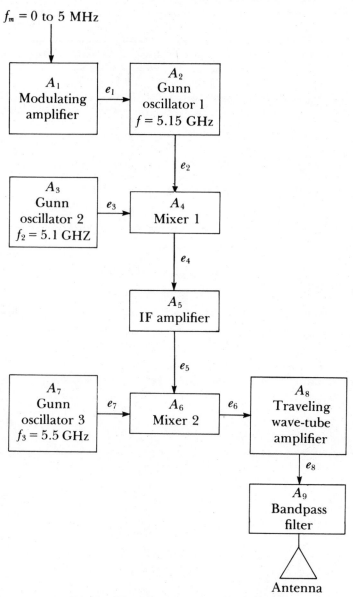

Fig. 4-57   Microwave transmitter block diagram.

the following characteristics:

Designation	Component	Characteristics
A_1	Modulating amplifier	$f_m = 0$ to 5 MHz
A_2	Gunn oscillator 1	$f_1 = 5.15$ GHz; modulation sensitivity is 0.25 MHz/V
A_3	Gunn oscillator 2	$f_2 = 5.1$ GHz
A_4	Mixer 1	$e_4 = a_1 e_1 + a_1 e_1^2$
A_5	IF amplifier	
A_6	Mixer 2	$e_6 = a_1 e_1 + a_1 e_1^2$
A_7	Gunn oscillator 3	$f_3 = 5.5$ GHz
A_8	Traveling-wave-tube amplifier	$A_v = K$ V/V
A_9	Bandpass filter	

FIND:

a. e_1, if the transmitter deviation is to be 2.5 MHz.

b. Give the frequencies present in the output e_4.

c. With no modulation, give the frequencies present at e_5 and e_6.

d. With modulation, what is the minimum bandwidth of the IF amplifier? What is the frequency range of interest?

e. With modulation, what should the center frequency and bandwidth be for the bandpass filter?

SOLUTION:

a. Since the modulation sensitivity of A_2 is 0.25 MHz/V, the magnitude of the voltage e_1 is

$$e_1 = 2.5 \text{ MHz} \times \frac{1 \text{ V}}{0.25 \text{ MHz}} = 10 \text{ V}$$

and our complete modulating waveform can be represented by:

$$e_1 = 10 \sin \omega_m t = 10 \sin (2\pi f_m t)$$

b. e_4 represents the output of mixer 1. With no modulation, we can represent e_2 and e_3 as follows:

$$e_2 = A_1 \sin \omega_1 t$$

$$e_3 = A_2 \sin \omega_2 t$$

and our output e_4 is represented by

$$e_4 = a_1(A_1 \sin \omega_1 t + A_2 \sin \omega_2 t) + a_1(A_1 \sin \omega_1 t + A_2 \sin \omega_2 t)^2$$

$$e_4 = a_1(A_1 \sin \omega_1 t + A_2 \sin \omega_2 t)$$
$$+ a_1[A_1^2 \sin^2 \omega_1 t + (2A_1A_2 \sin \omega_1 t)(\sin \omega_2 t) + A_2^2 \sin^2 \omega_2 t]$$

Using trigonometric identities, we find

$$e_4 = a_1(A_1 \sin \omega_1 t + A_2 \sin \omega_2 t) + a_1[0.5A_1^2(1 - \cos 2\omega_1 t)$$
$$+ 0.5A_2^2(1 - \cos 2\omega_2 t) + A_1A_2 \sin (\omega_1 + \omega_2)t$$
$$+ A_1A_2 \sin (\omega_1 - \omega_2)t]$$

Therefore the frequencies present are

Term	Frequency, GHz
ω_1	5.15
ω_2	5.10
$2\omega_1$	10.30
$2\omega_2$	10.20
$\omega_1 + \omega_2$	10.25
$\omega_1 - \omega_2$	0.05

c.   The selected signal out of the IF amplifier (e_5) is the difference frequency which contains the needed information. Of course, the lower frequency is selected because of the less stringent design procedures. Therefore f_0 is selected to be 0.05 GHz.

The signal out of the second mixer (e_6) can be determined as in step a, except the two frequencies of interest are $f_0 = 0.05$ GHz and $f_3 = 5.5$ GHz.

Term	Frequency, GHz
f_0	0.05
f_5	5.50
$2f_0$	0.10
$2f_5$	11.00
$f_5 + f_0$	5.55
$f_5 - f_0$	5.45

d. We can find our minimum bandwidth by finding our deviation ratio first:

$$\delta = \frac{f_d}{f_m} = \frac{2.5\text{ MHz}}{5\text{ MHz}} = 0.5$$

With this value of deviation ratio, we can look at the Bessel tables given in many engineering texts to determine the significant contributors to the output for the case when $\delta = 0.5$. The data of Table 4-14 is reconstructed for $\delta = 0.5$; let's assume that any contributions less than 0.001 are negligible. From this data, we can see that all spectral contributors up to and including $n = 3$ are significant.

TABLE 4-14

n	$J_n(0.5)$
0	0.9385
1	0.2423
2	0.0306
3	0.0026
4	0.0002

Therefore the minimum bandwidth can be determined to be:

$$\text{BW} = 4f_m = 4(5\text{ MHz}) = 20\text{ MHz}$$

This is centered at the IF of 50 MHz, so our range of interest is 40 to 60 MHz.

e. The assigned band is in the range of 5.5 to 6 GHz. Therefore, we would select a frequency in this range from part d. We will select $f_5 + f_0 = 5.55$ GHz. The minimum required bandwidth for the bandpass filter would be the same as for the IF stage, or 20 MHz.

CHECK THE SOLUTION: Check each derivation and calculation for accuracy.

APPENDICES

BIPOLAR TRANSISTOR GAIN
AND IMPEDANCE EQUATIONS

The small-signal equations shown in Appendices A1 through A6 can be used in the solution of most amplifier problems. In general, the voltage-gain equations and input-impedance equations are sufficient to permit most analyses. For each amplifier configuration, an ac equivalent circuit is drawn which has all power supplies returned to ac ground, and all reactive components replaced with their equivalent impedance at the frequency of interest (capacitors may be made short circuits or inductors may be made open circuits, for example). Also, for each amplifier configuration, a linear ac equivalent circuit is given whereby the transistor is replaced by its linear small-signal equivalent model and inserted into the ac equivalent circuit.

In all cases, approximate formulas are given and are immediately followed by the exact formulas and typical values. Only one parameter is usually calculated for the common-base, common-emitter, or common-collector amplifier, and that is h_{ib}, h_{ie}, or h_{ic}, respectively, each being a function of the dc emitter current. For example, $h_{ib} = (26\,\text{mV})/I_E$, which yields a resistance of 26 Ω with 1 mA of emitter current. The parameters h_{ie} and h_{ic} can also be calculated using the relationship $h_{ie} = h_{ic} = (h_{fe} + 1)h_{ib}$. The remainder of the parameters are usually given in a problem; if not, use the approximate values given by each appendix as typical values for each configuration, or neglect the parameters and set them equal to zero or infinity.

Of all the amplifier configurations, only one type gives signal inversion, and that is the common-emitter amplifier. Both the common-base and common-collector types are noninverting amplifiers. All amplifier configurations include a generator impedance because its magnitude is reflected in the output impedance calculations.

Appendix A1 provides the gain and impedance equations for the common-base (CB) amplifier, while Appendices A2 through A5 provide equations for the common emitter (CE), common collector (CC), common base with base resistor (CBR), and common emitter with emitter resistor (CER). Appendix A6 covers a special case often encountered in amplifier analysis and provides approximate gain and impedance equations for the CE collector-base-feedback (CBF) amplifier. Appendix A7 provides the h-parameter interrelationships for the CE, CB, and CC configurations, while Appendix A8 provides the hybrid-π equivalent circuit and its approximate interrelationships with the CE h-parameters.

Although all of the illustrations given in Appendix A are for *NPN*-type transistors, the equations are also applicable for *PNP* transistors. The techniques used to identify a particular configuration are as follows: If the input is on the emitter and the output is on the collector, the circuit is a CB amplifier; if the input is on the base and the output on the collector, it is a CE amplifier; and if the input is on the base and the output on the emitter it is a CC amplifier. In all cases, the common terminal describes the type of configuration.

APPENDIX A1
Common-Base (CB) Bipolar Amplifier Small-Signal Equations

Approximate formulas:

$$A_v = \frac{e_o}{e_i} \approx \frac{R_C}{h_{ib}} \qquad \text{(A1-1)}$$

$$Z_{in} \approx h_{ib} \qquad \text{(A1-2)}$$

$$A_i = \frac{i_c}{i_e} \approx h_{fb} \qquad \text{(A1-3)}$$

$$Z_o \approx \frac{h_{ib} + R_g}{h_{rb}} \qquad \text{(A1-4)}$$

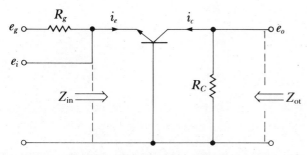

Fig. A1-1   CB ac equivalent circuit.

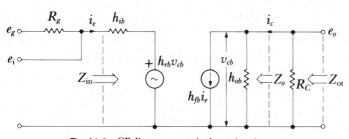

Fig. A1-2   CB linear ac equivalent circuit.

Exact formulas:

$$A_v = \frac{-h_{fb}R_C}{h_{ib} + R_C\,\Delta h_b} \tag{A1-5}$$

$$Z_{in} = \frac{h_{ib} + R_C\,\Delta h_b}{1 + R_C h_{ob}} \tag{A1-6}$$

$$A_i = \frac{h_{fb}}{1 + R_C h_{ob}} \tag{A1-7}$$

$$Z_o = \frac{h_{ib} + R_g}{\Delta h_b + R_g h_{ob}} \tag{A1-8}$$

$$Z_{ot} = R_C \| Z_o \tag{A1-9}$$

$$\Delta h_b = h_{ib} h_{ob} - h_{fb} h_{rb} \tag{A1-10}$$

Typical parameter values:

$$h_{ib} = \frac{0.026}{I_E} \qquad I_E = \text{dc emitter current} \tag{A1-11}$$

$$h_{ib} = 26\,\Omega \qquad \text{at } I_E = 1\,\text{mA}$$

$$h_{fb} = -0.98$$

$$h_{rb} = 3 \times 10^{-4}$$

$$h_{ob} = 5 \times 10^{-7}$$

APPENDIX A2
Common-Emitter (CE) Bipolar Amplifier Small-Signal Equations

Approximate formulas:

$$A_v = \frac{e_o}{e_i} \approx \frac{-R_C}{h_{ib}} \tag{A2-1}$$

$$Z_{in} \approx h_{ie} \tag{A2-2}$$

$$A_i = \frac{i_c}{i_b} \approx h_{fe} \tag{A2-3}$$

$$Z_o \approx \frac{1}{h_{oe}} \tag{A2-4}$$

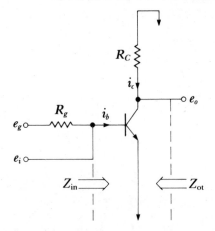

Fig. A2-1   CE ac equivalent circuit.

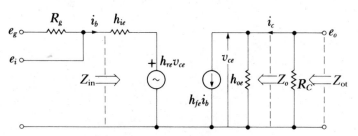

Fig. A2-2   CE linear ac equivalent circuit.

Exact formulas:

$$A_v = \frac{-h_{fe}R_C}{h_{ie} + R_C\,\Delta h_e} \tag{A2-5}$$

$$Z_{in} = \frac{h_{ie} + R_C\,\Delta h_e}{1 + R_C h_{oe}} \tag{A2-6}$$

$$A_i = \frac{h_{fe}}{1 + R_C h_{oe}} \tag{A2-7}$$

$$Z_o = \frac{h_{ie} + R_g}{\Delta h_e + R_g h_{oe}} \tag{A2-8}$$

$$Z_{ot} = R_C \| Z_o \tag{A2-9}$$

$$\Delta h_e = h_{ie}h_{oe} - h_{fe}h_{re} \tag{A2-10}$$

Typical parameter values:

$$h_{ib} = \frac{0.026}{I_E} \qquad I_E = \text{dc emitter current} \qquad \text{(A2-11)}$$

$$h_{fe} = 100$$

$$h_{oe} = 25 \times 10^{-6}$$

$$h_{re} = 25 \times 10^{-5}$$

$$h_{ie} = (h_{fe} + 1)h_{ib} \approx 2.6 \text{ k}\Omega \qquad \text{at } I_E = 1 \text{ mA} \qquad \text{(A2-12)}$$

APPENDIX A3
Common-Collector (CC) Bipolar Amplifier Small-Signal Equations

Approximate formulas:

$$A_v = \frac{e_o}{e_i} \approx \frac{R_E}{h_{ib} + R_E} \qquad \text{(A3-1)}$$

$$Z_{in} \approx -h_{fc}(h_{ib} + R_E) \qquad \text{(A3-2)}$$

$$A_i = \frac{i_e}{i_b} \approx h_{fc} \qquad \text{(A3-3)}$$

$$Z_o \approx \frac{h_{ic} + R_g}{-h_{fc}} \qquad \text{(A3-4)}$$

Exact formulas:

$$A_v = \frac{-h_{fc}R_E}{h_{ic} + R_E \, \Delta h_c} \qquad \text{(A3-5)}$$

$$Z_{in} = \frac{h_{ic} + R_E \Delta h_c}{1 + R_E h_{oc}} \qquad \text{(A3-6)}$$

$$A_i = \frac{h_{fc}}{1 + R_E h_{oc}} \qquad \text{(A3-7)}$$

$$Z_o = \frac{h_{ic} + R_g}{\Delta h_c + R_g h_{oc}} \qquad \text{(A3-8)}$$

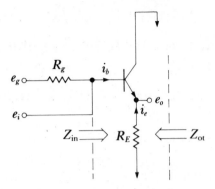

Fig. A3-1   CC ac equivalent circuit.

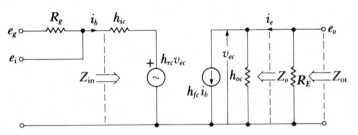

Fig. A3-2   CC linear ac equivalent circuit.

$$Z_{ot} = R_E \| Z_o \tag{A3-9}$$

$$\Delta h_e = h_{ic} h_{oc} - h_{fc} h_{rc} \tag{A3-10}$$

Typical parameter values:

$$h_{ib} = \frac{0.026}{I_E} \qquad \text{where } I_E = \text{dc emitter current} \tag{A3-11}$$

$$h_{ic} = (h_{fe} + 1)h_{ib} \approx 2.6 \text{ k}\Omega \qquad \text{at } I_E = 1 \text{ mA}$$

$$h_{oc} = 25 \times 10^{-6}$$

$$h_{fc} = -100$$

$$h_{rc} = 1$$

APPENDIX A4
Common-Base Amplifier with Base Resistor (CBR) Bipolar Small-Signal Equations

Approximate formulas:

$$A_v = \frac{e_o}{e_i} \approx \frac{R_C}{h_{ib} + \dfrac{R_B}{h_{fe}}} \qquad \text{(A4-1)}$$

$$Z_{in} \approx h_{ib} + \frac{R_B}{h_{fe}} \qquad \text{(A4-2)}$$

$$A_i = \frac{i_c}{i_e} \approx h_{fb} \qquad \text{(A4-3)}$$

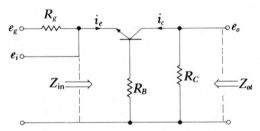

Fig. A4-1   CBR ac equivalent circuit.

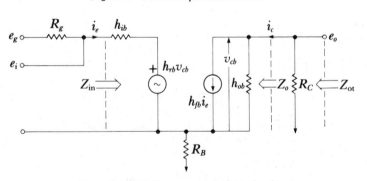

Fig. A4-2   CBR linear ac equivalent circuit.

$$Z_o \approx \frac{h_{ib} + R_g + R_B(1 + h_{fb})}{\Delta h_b' + R_g h_{ob}} \tag{A4-4}$$

Exact formulas:

$$A_v = \frac{(h_{ob}R_B - h_{fb})R_C}{(1 - h_{rb})(1 + h_{fb})R_B + (h_{ib} + R_C\,\Delta h_b')(1 + h_{ob}R_B)} \tag{A4-5}$$

$$Z_{in} = \frac{(1 - h_{rb})(1 + h_{fb})R_B + (h_{ib} + R_C\Delta h_b')(1 + h_{ob}R_B)}{1 + h_{ob}(R_B + R_C)} \tag{A4-6}$$

$$A_i = \frac{h_{fb} - h_{ob}R_B}{1 + h_{ob}(R_B + R_C)} \tag{A4-7}$$

$$Z_o = \frac{(h_{ib} + R_g)(1 + h_{ob}R_B) + (1 - h_{rb})(1 + h_{fb})R_B}{\Delta h_b'(1 + h_{ob}R_B) + R_g h_{ob}} \tag{A4-8}$$

$$Z_{ot} = Z_o \| R_C \tag{A4-9}$$

$$\Delta h_b' = \frac{[h_{ib}(1 + h_{ob}R_B) + (1 - h_{rb})(1 + h_{fb})R_B]h_{ob} - (h_{rb} + h_{ob}R_B)(h_{fb} - h_{ob}R_B)}{(1 + h_{ob}R_B)^2} \tag{A4-10}$$

Typical parameter values:

$$h_{ib} = \frac{0.026}{I_E} \qquad I_E = \text{dc emitter current} \tag{A4-11}$$

$$h_{ib} = 26\ \Omega \qquad \text{at } I_E = 1\ \text{mA}$$
$$h_{fb} = -0.98$$
$$h_{rb} = 3 \times 10^{-4}$$
$$h_{ob} = 5 \times 10^{-7}$$

APPENDIX A5
Common-Emitter Amplifier
with Emitter Resistor (CER)
Small-Signal Equations

Approximate formulas:

$$A_v = \frac{e_o}{e_i} \approx \frac{-R_C}{h_{ib} + R_E} \tag{A5-1}$$

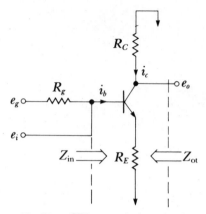

Fig. A5-1   CER ac equivalent circuit.

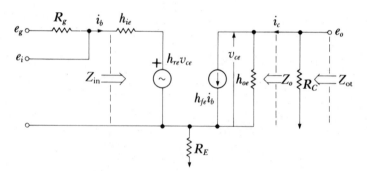

Fig. A5-2   CER linear ac equivalent circuit.

$$Z_{in} \approx h_{fe}(h_{ib} + R_E) \qquad \text{(A5-2)}$$

$$A_i = \frac{i_c}{i_b} \approx h_{fe} \qquad \text{(A5-3)}$$

$$Z_o \approx \frac{1}{h_{oe}} \qquad \text{(A5-4)}$$

Exact formulas:

$$A_v = \frac{(h_{oe}R_E - h_{fe})R_C}{(h_{ie} + R_C \,\Delta h_e')(1 + h_{oe}R_E) + (1 - h_{re})(1 + h_{fe})R_E} \qquad \text{(A5-5)}$$

$$Z_{in} = \frac{(1 + h_{oe}R_E)(h_{ie} + R_C\,\Delta h'_e) + (1 - h_{re})(1 + h_{fe})R_E}{1 + (R_C + R_E)h_{oe}} \tag{A5-6}$$

$$A_i = \frac{h_{fe} - h_{oe}R_E}{1 + (R_C + R_E)h_{oe}} \tag{A5-7}$$

$$Z_o = \frac{(h_{ie} + R_g)(1 + h_{oe}R_E) + (1 - h_{re})(1 + h_{fe})R_E}{\Delta h'_e(1 + h_{oe}R_E) + R_g h_{oe}} \tag{A5-8}$$

$$Z_{ot} = Z_o \| R_C \tag{A5-9}$$

$$\Delta h'_e = \frac{[h_{ie}(1 + h_{oe}R_E) + (1 - h_{re})(1 + h_{fe})R_E]h_{oe} - (h_{re} + h_{oe}R_E)(h_{fe} - h_{oe}R_E)}{(1 + h_{oe}R_E)^2} \tag{A5-10}$$

Typical parameter values:

$$h_{ib} = \frac{0.026}{I_E} \qquad I_E = \text{dc emitter current} \tag{A5-11}$$

$$h_{fe} = 100$$
$$h_{oe} = 25 \times 10^{-6}$$
$$h_{re} = 25 \times 10^{-5}$$
$$h_{ie} = (h_{fe} + 1)h_{ib} = 2.6\ \text{k}\Omega \qquad \text{at 1 mA} \tag{A5-12}$$

APPENDIX A6
Collector-Base Feedback (CBF) Bipolar Amplifier Small-Signal Equations

Approximate equations:

$$A_v = \frac{e_o}{e_i} \approx \frac{-h_{fe}^* R_L}{h_{ie}^*(h_{ie}^*/R_F + 1 + h_{oe}^* R_L)} \approx \frac{-h_{fe}^* R_L}{h_{ie}^*} \tag{A6-1}$$

$$Z_{in} \approx \frac{R_F h_{ie}^*}{R_F + h_{fe}^* R_L} \tag{A6-2}$$

$$A_i = \frac{i_c^*}{i_b^*} \approx \frac{h_{fe}^*}{1 + h_{fe}^* R_L/R_F} \approx \frac{R_F}{R_L} \tag{A6-3}$$

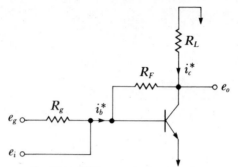

Fig. A6-1   CBF ac equivalent circuit.

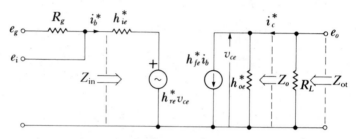

Fig. A6-2   CBF linear ac equivalent circuit.

$$Z_o \approx \frac{(h_{ie}^* + R_g)R_F}{(h_{ie}^* + R_g)h_{oe}^* R_F + R_g h_{fe}^*} \tag{A6-4}$$

Exact equations:

$$A_v = \frac{-h_{fe}^* R_L}{\Delta h^* R_L + h_{ie}^*} \tag{A6-5}$$

$$Z_{in} = \frac{\Delta h^* R_L + h_{ie}^*}{1 + h_{oe}^* R_L} \tag{A6-6}$$

$$A_i = \frac{h_{fe}^*}{1 + h_{oe}^* R_L} \tag{A6-7}$$

$$Z_o = \frac{h_{ie}^* + R_g}{\Delta h^* + h_{oe}^* R_g} \tag{A6-8}$$

$$Z_{ot} = Z_o \| R_L \tag{A6-9}$$

* Equations:

$$h_{ie}^* = \frac{h_{ie}R_F}{h_{ie} + R_F} \tag{A6-10}$$

$$h_{oe}^* = h_{oe} + \frac{(1 - h_{re})(1 + h_{fe})}{h_{ie} + R_F} \tag{A6-11}$$

$$h_{fe}^* = \frac{h_{fe}R_F - h_{ie}}{h_{ie} + R_F} \tag{A6-12}$$

$$h_{re}^* = h_{re} + \frac{(1 - h_{re})h_{ie}}{h_{ie} + R_F} \tag{A6-13}$$

$$\Delta h^* = h_{ie}^* h_{oe}^* - h_{fe}^* h_{re}^* \tag{A6-14}$$

APPENDIX A7
h-Parameter
Interrelationships

CB = f(CE):

$$h_{ib} = \frac{h_{ie}}{(1 + h_{fe})(1 - h_{re}) + h_{ie}h_{oe}} \approx \frac{h_{ie}}{1 + h_{fe}} \tag{A7-1}$$

$$h_{ob} = \frac{h_{oe}}{(1 + h_{fe})(1 - h_{re}) + h_{ie}h_{oe}} \approx \frac{h_{oe}}{1 + h_{fe}} \tag{A7-2}$$

$$h_{fb} = \frac{-h_{fe}(1 - h_{re}) - h_{ie}h_{oe}}{(1 + h_{fe})(1 - h_{re}) + h_{ie}h_{oe}} \approx \frac{-h_{fe}}{1 + h_{fe}} \tag{A7-3}$$

$$h_{rb} = \frac{h_{ie}h_{oe} - h_{re}(1 + h_{fe})}{(1 + h_{fe})(1 - h_{re}) + h_{ie}h_{oe}} \approx \frac{h_{ie}h_{oe}}{1 + h_{fe}} - h_{re} \tag{A7-4}$$

CB = f(CC):

$$h_{ib} = \frac{h_{ic}}{h_{ic}h_{oc} - h_{fc}h_{rc}} \approx \frac{-h_{ic}}{h_{fc}} \tag{A7-5}$$

$$h_{ob} = \frac{h_{oc}}{h_{ic}h_{oc} - h_{fc}h_{rc}} \approx \frac{h_{oc}}{h_{fc}} \tag{A7-6}$$

$$h_{fb} = \frac{h_{rc}(1 + h_{fc}) - h_{ic}h_{oc}}{h_{ic}h_{oc} - h_{fc}h_{rc}} \approx \frac{-(1 + h_{fc})}{h_{fc}} \tag{A7-7}$$

$$h_{rb} = \frac{h_{fc}(1 - h_{rc}) + h_{ic}h_{oc}}{h_{ic}h_{oc} - h_{fc}h_{rc}} \approx h_{rc} - 1 - \frac{h_{ic}h_{oc}}{h_{fc}} \qquad \text{(A7-8)}$$

CE = f(CB):

$$h_{ie} = \frac{h_{ib}}{(1 + h_{fb})(1 - h_{rb}) + h_{ob}h_{ib}} \approx \frac{h_{ib}}{1 + h_{fb}} \qquad \text{(A7-9)}$$

$$h_{oe} = \frac{h_{ob}}{(1 + h_{fb})(1 - h_{rb}) + h_{ob}h_{ib}} \approx \frac{h_{ob}}{1 + h_{fb}} \qquad \text{(A7-10)}$$

$$h_{fe} = \frac{-h_{fb}(1 - h_{rb}) - h_{ob}h_{ib}}{(1 + h_{fb})(1 - h_{rb}) + h_{ob}h_{ib}} \approx \frac{-h_{fb}}{1 + h_{fb}} \qquad \text{(A7-11)}$$

$$h_{re} = \frac{h_{ib}h_{ob} - h_{rb}(1 + h_{fb})}{(1 + h f_b)(1 - h_{rb}) + h_{ob}h_{ib}} \approx \frac{h_{ib}h_{ob}}{1 + h_{fb}} - h_{rb} \qquad \text{(A7-12)}$$

CE = f(CC):

$$h_{ie} = h_{ic} \qquad \text{(A7-13)}$$

$$h_{oe} = h_{oc} \qquad \text{(A7-14)}$$

$$h_{fe} = -(1 + h_{fc}) \qquad \text{(A7-15)}$$

$$h_{re} = 1 - h_{rc} \qquad \text{(A7-16)}$$

CC = f(CB):

$$h_{ic} = \frac{h_{ib}}{(1 + h_{fb})(1 - h_{rb}) + h_{ob}h_{ib}} \approx \frac{h_{ib}}{1 + h_{fb}} \qquad \text{(A7-17)}$$

$$h_{oc} = \frac{h_{ob}}{(1 + h_{fb})(1 - h_{rb}) + h_{ob}h_{ib}} \approx \frac{h_{ob}}{1 + h_{fb}} \qquad \text{(A7-18)}$$

$$h_{fc} = \frac{h_{rb} - 1}{(1 + h_{fb})(1 - h_{rb}) + h_{ob}h_{ib}} \approx \frac{-1}{1 + h_{fb}} \qquad \text{(A7-19)}$$

$$h_{rc} = \frac{1 + h_{fb}}{(1 + h_{fb})(1 - h_{rb}) + h_{ob}h_{ib}} \approx 1 \qquad \text{(A7-20)}$$

CC = f(CE):

$$h_{ic} = h_{ie} \qquad \text{(A7-21)}$$

$$h_{oc} = h_{oe} \qquad \text{(A7-22)}$$

$$h_{fc} = -(1 + h_{fe}) \qquad \text{(A7-23)}$$

$$h_{rc} = 1 - h_{re} \qquad \text{(A7-24)}$$

APPENDIX A8
Hybrid-π Bipolar
Small-Signal Equivalent

The hybrid-π model is another model that represents the CE amplifier with frequency-dependent parameters included. In particular, the high-frequency roll-off of h_{fe} is represented by C_π while the intrinsic collector-base capacitance is represented by C_μ. The hybrid-π model is shown with the approximate CE h-parameter equations in Fig. A8-1.

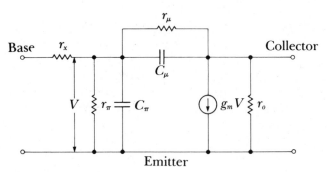

$$g_m = \frac{1}{h_{ib}} \tag{A8-1}$$

$$r_\pi = h_{fe} h_{ib} \tag{A8-2}$$

$$r_x \approx h_{ie} - r_\pi \text{ at 1 kHz (typical value = 10 to 50 } \Omega) \tag{A8-3}$$

$$r_\mu \approx \frac{r_\pi}{h_{re}} \text{ at 1 kHz} \tag{A8-4}$$

$$1/r_o \approx h_{oe} - g_m h_{re} \tag{A8-5}$$

$$C_\mu = C_{ob} - C_{bc} - C_{ce} \tag{A8-6}$$

$$C_\pi = \frac{g_m}{\omega_T} - C_\mu \approx \frac{g_m}{\omega_T} \tag{A8-7}$$

$$\omega_T = 2\pi f_T \ (f_T = \text{CB high-frequency cutoff}) \tag{A8-8}$$

Fig. A8-1   Hybrid-π transistor representation.

B

JFET SMALL-SIGNAL
EQUATIONS

The small-signal gain and impedance formulas for the JFET are given in Appendices B1 through B3 for the common source, common gate, and common drain, respectively. As was the case in Appendix A for the bipolar transistor, the input-impedance and voltage-gain equations are about all that is needed to analyze almost any amplifier problem. As described in Appendix A, the ac equivalent and linear ac equivalent circuits are given for each configuration.

In all cases, the exact formulas are given immediately and are followed by the approximation formulas and typical parameter values. As can be seen by the voltage-gain equations, only one configuration gives signal inversion and that is the common-source amplifier.

APPENDIX B1
JFET Common-Source (CS) Small-Signal Equations

Exact formulas:

$$A_v = \frac{e_o}{e_i} = \frac{-g_{fs}R_D}{1 + g_o(R_D + R_S) + g_{fs}R_S} \tag{B1-1}$$

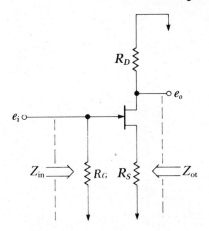

Fig. B1-1   CS ac equivalent circuit.

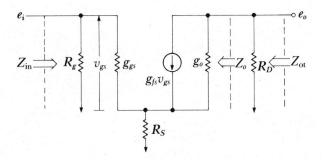

Fig. B1-2   CS linear ac equivalent circuit.

$$Z_{in} = \left[\frac{1/g_{gs}}{1 - \dfrac{g_{fs}R_S}{1 + g_o(R_D + R_S) + g_{fs}R_S}} \right] \| R_G \qquad (B1\text{-}2)$$

$$Z_o = \frac{(1 + g_{fs}R_S)R_D}{g_o(R_D + R_S)} \qquad (B1\text{-}3)$$

$$Z_{ot} = Z_o \| R_D \qquad (B1\text{-}4)$$

Approximate formulas:

$$A_v \approx \frac{-g_{fs}R_D}{1 + g_{fs}R_S} \qquad (B1\text{-}5)$$

$$Z_{in} \approx R_G \qquad \text{(B1-6)}$$

$$Z_{ot} \approx R_D \qquad \text{(B1-7)}$$

Typical parameter values:

$$g_{fs} = 4 \times 10^{-3}$$
$$g_o = 5 \times 10^{-5}$$
$$g_{gs} = 10^{-9}$$

APPENDIX B2
JFET Common-Gate (CG) Small-Signal Equations

Exact formulas:

$$A_v = \frac{e_o}{e_i} = \frac{g_{fs}R_D}{1 + g_o(R_D + R_X) + g_{fs}R_X} \qquad \text{(B2-1)}$$

$$Z_{in} = \frac{1}{g_o + g_{fs}} + \frac{(R_X + R_D)g_o}{g_o + g_{fs}} + \frac{g_{fs}R_X}{g_o + g_{fs}} \qquad \text{(B2-2)}$$

$$Z_o = \frac{(1 + g_{fs}R_S)R_D}{g_o(R_D + R_S)} \qquad \text{(B2-3)}$$

$$Z_{ot} = Z_o \| R_D \qquad \text{(B2-4)}$$

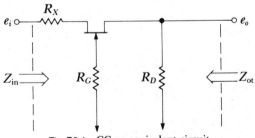

Fig. B2-1   CG ac equivalent circuit.

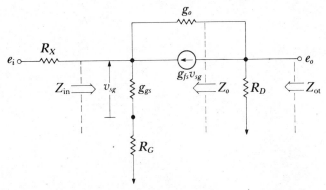

Fig. B2-2   CG linear ac equivalent circuit.

Approximate formulas:

$$A_v \approx \frac{g_{fs}R_D}{1 + g_{fs}R_X}$$ (B2-5)

$$Z_{in} \approx R_X + 1/g_{fs}$$ (B2-6)

$$Z_{ot} \approx R_D$$ (B2-7)

Typical parameter values:

$$g_{fs} = 4 \times 10^{-3}$$
$$g_o = 5 \times 10^{-5}$$
$$g_{gs} = 10^{-9}$$

APPENDIX B3
JFET Common-Drain (CD) Small-Signal Equations

Exact formulas:

$$A_v = \frac{e_o}{e_i} = \frac{g_{fs}R_S}{1 + R_S(g_{fs} + g_o)}$$ (B3-1)

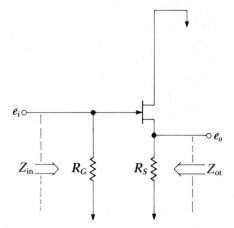

Fig. B3-1   CD ac equivalent circuit.

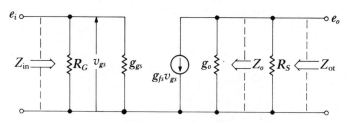

Fig. B3-2   CD linear ac equivalent circuit.

$$Z_{\text{in}} = \left\{ \dfrac{1/g_{gs}}{1 - \left[\dfrac{g_{fs}R_S}{1 + R_S(g_{fs} + g_o)} \right]} \right\} \Big\| R_G = \left(\dfrac{1/g_{gs}}{1 - A_v} \right) \Big\| R_G \qquad (B3\text{-}2)$$

$$Z_o = \dfrac{1 + g_o R_S}{g_{fs}} \qquad (B3\text{-}3)$$

$$Z_{ot} = R_S \| Z_o \qquad (B3\text{-}4)$$

Approximate formulas:

$$A_v \approx \dfrac{-g_{fs}R_S}{1 + g_{fs}R_S} \qquad (B3\text{-}5)$$

$$Z_{\text{in}} \approx R_G \qquad (B3\text{-}6)$$

$$Z_{ot} \approx R_S \| (1/g_{fs}) \qquad (B3\text{-}7)$$

Typical parameter values:

$$g_{fs} = 4 \times 10^{-3}$$
$$g_o = 5 \times 10^{-5}$$
$$g_{gs} = 10^{-9}$$

OPERATIONAL AMPLIFIER EQUATIONS

Appendix C provides equations for use in the analysis of operational amplifiers. Appendix C1 provides the small-signal ac equations for the noninverting configuration, and Appendix C2 provides the small-signal ac equations for the inverting configuration. Appendix C3 describes a technique for the dc analysis of operational amplifiers and provides a generalized equation that may be used in the solution of many dc amplifier problems.

APPENDIX C1*
Noninverting Operational Amplifier Equations

Figure C1-1 shows a noninverting amplifier configuration, where

Z_{in} = input impedance of total amplifier
Z_o = output impedance of total amplifier

*Equations taken from material presented in "RCA Linear Integrated Circuits," pp. 173–177, copyright 1974 by RCA Corporation.

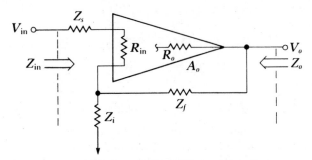

Fig. C1-1   Noninverting amplifier configuration.

A_o = open-loop gain of operational amplifier (A_o is positive)
R_o = output resistance of operational amplifier
R_{in} = input resistance of operational amplifier

Gain:

$$A_v = \frac{V_o}{V_{in}} \approx \frac{Z_f + Z_i}{Z_i} \qquad \text{(for } A_o \text{ and } R_{in} \text{ large, } R_o \text{ small)} \qquad \text{(C1-1)}$$

$$= \frac{Z_i R_o + A_o R_{in}(Z_i + Z_f)}{(R_{in} + Z_s)(Z_i + Z_f + R_o) + Z_i(Z_f + R_o) + A_o R_{in} Z_i} \qquad \text{(C1-2)}$$

Input impedance:

$$Z_{in} \approx \frac{A_o R_{in} Z_i}{Z_i + Z_f} \qquad \text{(for } A_o \text{ and } R_{in} \text{ large, } R_o \text{ small)} \qquad \text{(C1-3)}$$

$$= Z_s + R_{in} + \frac{Z_i(Z_f + R_o + A_o R_{in})}{Z_i + Z_f + R_o} \qquad \text{(C1-4)}$$

Output impedance:

$$Z_o \approx \frac{R_o(Z_i + Z_f)}{A_o Z_i} \qquad \text{(for } A_o \text{ and } R_{in} \text{ large)} \qquad \text{(C1-5)}$$

$$= \frac{R_o[(Z_i + Z_f)(Z_s + R_{in}) + Z_i Z_f][Z_i R_o + A_o R_{in}(Z_i + Z_f)]}{[A_o R_{in}(Z_i + Z_f)][(R_{in} + Z_s)(Z_i + Z_f + R_o) + Z_i(Z_f + R_o) + A_o R_{in} Z_i]}$$
$$\text{(C1-6)}$$

APPENDIX C2*
Inverting Operational Amplifier Equations

Figure C2-1 illustrates an inverting amplifier configuration, where

Z_{in} = input impedance of total amplifier
Z_o = output impedance of total amplifier
A_o = open-loop gain of operational amplifier (A_o is negative)
R_o = output resistance of operational amplifier
R_{in} = input resistance of operational amplifier

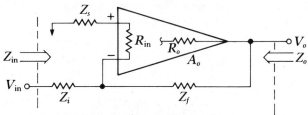

Fig. C2-1   Inverting amplifier configuration.

Gain:

$$A_v = \frac{V_o}{V_{in}} \approx \frac{-Z_f}{Z_i} \quad \text{(for } A_o \text{ and } R_{in} \text{ large, } R_o \text{ small)} \tag{C2-1}$$

$$= \frac{V_o}{V_{in}} = \frac{R_o(R_{in} + Z_s) + A_o R_{in} Z_f}{(Z_f + R_o)(R_{in} + Z_s) + Z_i(Z_f + R_o + R_{in} + Z_s) - A_o R_{in} Z_i} \tag{C2-2}$$

Input impedance:

$$Z_{in} \approx Z_i \quad \text{(for } A_o \text{ and } R_{in} \text{ large, } R_o \text{ small)} \tag{C2-3}$$

$$= Z_i + \frac{(Z_f + R_o)(R_{in} + Z_s)}{(Z_f + R_o + R_{in} + Z_s) - A_o R_{in}} \tag{C2-4}$$

Output impedance:

$$Z_o \approx \frac{R_o(Z_f + Z_i)}{-Z_i A_o} \quad \text{(for } R_{in} \text{ large)} \tag{C2-5}$$

$$= \frac{R_o[Z_i(Z_f + R_{in} + Z_s) + Z_f(R_{in} + Z_s)]}{-A_o R_{in} Z_i} \tag{C2-6}$$

*Equations taken from material presented in "RCA Linear Integrated Circuits," pp. 177–180, copyright 1974 by RCA Corporation.

APPENDIX C3
DC Analysis of
Operational Amplifiers

DC analysis of operational amplifiers can be simplified if analysis begins at the positive terminal, which is correctly designated to be the independent forcing function. The circuit of Fig. C3-1 can be used in

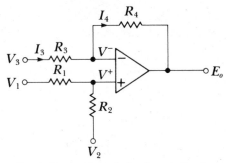

Fig. C3-1   Circuit for the dc analysis of operational amplifiers.

the analysis of most operational amplifiers. The defining equations are:

$$V^+ = \frac{(V_1 - V_2)R_2}{R_1 + R_2} + V_2$$

$V^+ = V^-$ (this is true for all ideal operational amplifiers)

$$I_3 = \frac{V_3 - V^-}{R_3}$$

$$I_3 = I_4$$

$$E_o = V^- - I_4 R_4$$

Solving, we find

$$E_o = \frac{(V_1 - V_2)(R_3 + R_4)R_2}{(R_1 + R_2)R_3} + \frac{(R_3 + R_4)V_2}{R_3} - \frac{V_3 R_4}{R_3} \qquad (C3\text{-}1)$$

EXAMPLE 1

Assume we are to find E_o for the circuit of Fig. C3-2. For the terms in

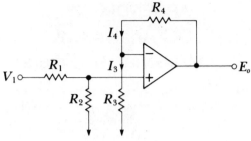

$$V_1 = 5 \text{ V dc} \qquad R_3 = 3 \text{ k}\Omega$$
$$R_1 = 1 \text{ k}\Omega \qquad R_4 = 4 \text{ k}\Omega$$
$$R_2 = 2 \text{ k}\Omega$$

Fig. C3-2   Amplifier circuit used in Example 1.

Eq. (C3-1), $V_2 = V_3 = 0$ V dc. Substituting into Eq. (C3-1), we find

$$E_o = \frac{(5-0)(2 \times 10^3)(7 \times 10^3)}{(1 \times 10^3 + 2 \times 10^3)3 \times 10^3} = 7.78 \text{ V dc}$$

$$= 7.78 \text{ V dc}$$

CHECK:

$$V^+ = \frac{R_2 V_1}{R_1 + R_2} = 3.333 \text{ V dc}$$

$$V^- = V^+ = 3.333 \text{ V dc}$$

$$I_3 = I_4 = \frac{V^-}{R_3} = \frac{3.333}{3 \times 10^3} = 1.111 \text{ mA}$$

$$E_o = V^- + I_o R_4 = 3.333 + (1.111 \text{ mA})(4 \text{ k}\Omega) = 7.78 \text{ V dc}$$

and this checks.

EXAMPLE 2 _____

GIVEN: The circuit of Fig. C3-3.

FIND: E_o.

For the circuit of Fig. C3-1, we have the following values determined

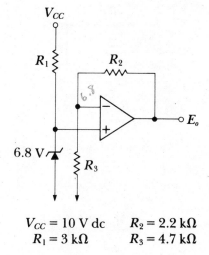

$$V_{CC} = 10 \text{ V dc} \qquad R_2 = 2.2 \text{ k}\Omega$$
$$R_1 = 3 \text{ k}\Omega \qquad R_3 = 4.7 \text{ k}\Omega$$

Fig. C3-3 Amplifier circuit used in Example 2.

by Fig. C3-3:

$$R_1 = 0 \qquad V_1 = 6.8 \text{ V} \qquad V_3 = 0 \text{ V} \qquad R_4 = 2.2 \text{ k}\Omega$$
$$R_2 = \infty \qquad V_2 = 0 \text{ V} \qquad R_3 = 4.7 \text{ k}\Omega$$

$$E_o = \left[\frac{(6.8 - 0)}{0/\infty + 1} \right] \frac{6.9 \times 10^3}{4.7 \times 10^3} + 0 - 0 \qquad \text{(C3-1)}$$

$$= 9.983 \text{ V dc}$$

Check this solution as an exercise.

APPENDIX C4
Offset Analysis Of
Operational Amplifiers

V_{ost} = offset at output of amplifier
I_{os} = specified offset current $(I_{B1} - I_{B2})$
V_{os} = specified offset voltage

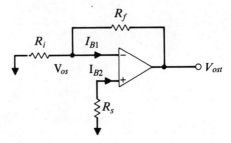

Fig. C4-1   Circuit for the dc offset analysis of operational amplifiers.

I_B = specified base current $[(I_{B1} + I_{B2})/2]$

$R_{eq} = R_i \| R_f$

$$V_{ost} = \left[V_{os} + I_B(R_s - R_{eq}) + \frac{I_{os}(R_s + R_{eq})}{2} \right] \frac{R_f + R_i}{R_i} \quad (C4\text{-}1)$$

MISCELLANEOUS
CIRCUIT EQUATIONS

APPENDIX D1
Delta-Y
Transformation

A transformation exists that relates the impedances of a delta connection
to that of a Y connection as follows (see Fig. D1-1):

$$Z_1 = \frac{Z_A Z_B}{Z_A + Z_B + Z_C} \qquad \text{(D1-1)}$$

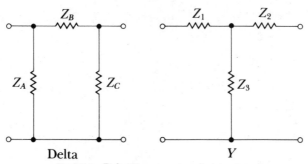

Delta   Y

Fig. D1-1   Delta-Y conversion designation.

$$Z_2 = \frac{Z_B Z_C}{Z_A + Z_B + Z_C} \tag{D1-2}$$

$$Z_3 = \frac{Z_A Z_C}{Z_A + Z_B + Z_C} \tag{D1-3}$$

$$Z_A = \frac{Z_1 Z_2 + Z_2 Z_3 + Z_1 Z_3}{Z_2} \tag{D1-4}$$

$$Z_B = \frac{Z_1 Z_2 + Z_2 Z_3 + Z_1 Z_3}{Z_3} \tag{D1-5}$$

$$Z_C = \frac{Z_1 Z_2 + Z_2 Z_3 + Z_1 Z_3}{Z_1} \tag{D1-6}$$

APPENDIX D2
RC and RL
Rise-Time Equations

There are several two-element networks that are used repeatedly in network analysis when a step voltage is applied (Figs. D2-1 through D2-5). Several of these are given below with the defining equations, V being the peak magnitude of the step input. Initial currents and charges are assumed to be zero for inductors and capacitors respectively except for the circuit of Fig. D2-5.

$$i(t) = \frac{V}{R}(1 - e^{-(R/L)t}) \tag{D2-1}$$

$$e_o(t) = V(e^{-(R/L)t}) \tag{D2-2}$$

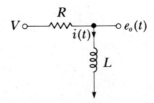

Fig. D2-1 RL high-pass circuit.

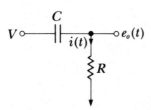

Fig. D2-2   *RL* low-pass circuit.

$$i(t) = \frac{V}{R}(1 - e^{-(R/L)t}) \qquad \text{(D2-3)}$$

$$e_o(t) = V(1 - e^{-(R/L)t}) \qquad \text{(D2-4)}$$

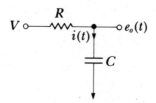

Fig. D2-3   *RC* high-pass circuit.

$$i(t) = \frac{V}{R}(e^{-t/RC}) \qquad \text{(D2-5)}$$

$$e_o(t) = V(e^{-t/RC}) \qquad \text{(D2-6)}$$

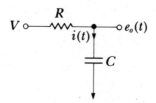

Fig. D2-4   *RC* low-pass circuit.

$$i(t) = \frac{Ve^{-t/RC}}{R} \qquad \text{(D2-7)}$$

$$e_o(t) = V(1 - e^{-t/RC}) \qquad \text{(D2-8)}$$

There is a configuration (a variation of the circuit shown in Fig. D2-4) that includes an initial charge on the capacitor. This expression is used quite extensively in network analysis.

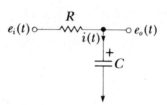

Fig. D2-5   *RC* circuit with an initial charge.

$$q(0^+) = CV_I$$

V_I = initial voltage on the capacitor

$e_i(t = \infty) = V_F$ = the final voltage on the capacitor; peak value of the step input.

$$e_i(s) = \frac{V_F}{s}$$

$$= \frac{1}{Cs}[I(s) + q(0^+)] + I(s)R$$

$$\frac{V_F}{s} = \frac{1}{Cs}[I(s) + CV_I] + I(s)R$$

$$\frac{V_F - V_I}{s} = \left(\frac{RCs + 1}{Cs}\right)I(s)$$

$$I(s) = \left(\frac{V_F - V_I}{RCs + 1}\right)\frac{C}{1} = \frac{1}{R}\left(\frac{V_F - V_I}{s + 1/RC}\right)$$

$$e_o(s) = I(s)\left(\frac{1}{Cs}\right) + V_I$$

$$= \frac{1}{RC}\left[\frac{V_F - V_I}{s(s + 1/RC)}\right] + V_I$$

$$e_o(t) = V_I + (V_F - V_I)(1 - e^{-t/RC}) \qquad \text{(D2-9)}$$

or $\qquad e_o(t) = V_F + (V_I - V_F)e^{-t/RC} \qquad \text{(D2-10)}$

APPENDIX D3
Transformer Equations

The schematic for a transformer is shown in Fig. D3-1.

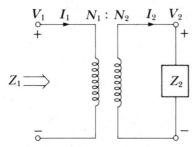

N_1 = primary turns
N_2 = secondary turns

Fig. D3-1   Transformer schematic.

$$\frac{V_1}{V_2} = \frac{N_1}{N_2} = n \qquad (n = \text{turns ratio}) \qquad \text{(D3-1)}$$

$$P_1 = P_2 \qquad \text{(ideally)}$$

$$\frac{I_1}{I_2} = \frac{N_2}{N_1} = \frac{1}{n} \qquad\qquad \text{(D3-2)}$$

$$Z_1 = n^2 Z_2 \qquad\qquad \text{(D3-3)}$$

APPENDIX D4
Multipole
Bandwidth Equations

There exists a simple relationship that relates the overall 3-dB band-width ($f_{3\,dB}$) for a multipole function with identically located poles. If N represents the number of poles and f_H equals the upper 3-dB frequency of each of the poles, then the overall bandwidth is

$$f_{3\,\mathrm{dB}H} = f_H (2^{1/N} - 1)^{1/2} \qquad (\text{D4-1})$$

Similarly, for an Nth-order lower-frequency breakpoint:

$$f_{3\,\mathrm{dB}L} = \frac{f_L}{(2^{1/N} - 1)^{1/2}} \qquad (\text{D4-2})$$

The midband frequency can be determined from:

$$f_{\mathrm{mid}} = (f_L f_H)^{1/2} \qquad (\text{D4-3})$$

or for Nth order:

$$f_{\mathrm{mid}} = (f_{3\mathrm{dB}H} f_{3\mathrm{dB}L})^{1/2} \qquad (\text{D4-4})$$

LAPLACE TRANSFORM
TABLE OF FORCING FUNCTIONS

TABLE E-1

(E1-1)	$\dfrac{\tanh (as/2)}{s}$	
(E1-2)	$\dfrac{E}{a}\left[\dfrac{1}{s^2} - \dfrac{a}{2s}\coth\left(\dfrac{as}{2} - 1\right)\right]$	
(E1-3)	$\dfrac{1 + \coth (as/2)}{2s}$	
(E1-4)	$\dfrac{aE}{(s^2 + a^2)(1 - e^{-\pi s/a})}$	
(E1-5)	$\dfrac{aE}{s^2 + a^2} - E\coth (\pi s/2a)$	
(E1-6)	$\dfrac{E\tanh (as/2)}{s^2}$	

275

(E1-7)	$\dfrac{E}{s(1 - e^{-as})}$	Square wave pulses at $0, a, 2a, 3a, 4a, 5a \ \dots t$ (amplitude E)
(E1-8)	$\dfrac{E(1 - e^{-as})}{s(1 - e^{-Ts})}$	Pulses from a to T, $T+a$ to $2T$, $2T+a \dots t$ (amplitude E)
(E1-9)	$\dfrac{E[1 - (s + a)e^{-T/2}]}{(s + a)(1 - e^{-Ts})}$	Decayed exponential at $0, \dfrac{T}{2}, T, \dfrac{3T}{2}, 2T, \dfrac{5T}{2} \dots t$ (Decayed exponential)
(E1-10)	$\dfrac{1}{as^2} - \dfrac{e^{-as}}{s(1 - e^{-as})}$	Sawtooth wave at $a, 2a, 3a, 4a, 5a \dots t$ (amplitude E)
(E1-11)	$\dfrac{E}{as^2} - \dfrac{Ee^{-as}}{s(1 - e^{-as})}$ or $\dfrac{E\tanh(as/2)}{s}$	Square wave between $+E$ and $-E$ at $a, 2a, 3a, 4a, 5a \dots t$

276

JFET DC AND AC
PARAMETER EQUATIONS

The ac and dc parameters are closely related for the JFET. The analysis will be broken into two parts: one part dealing with operation in the pinch-off region (region where I_D is fairly constant with increasing drain-to-source voltage V_{DS}), and the other part dealing with operation in the triode region (region from $V_{DS} = 0$ up to pinch-off). These two regions are shown in Fig. F-1. Usually, JFETs are operated

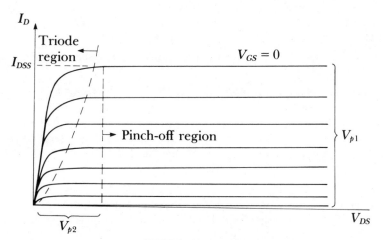

Fig. F-1   JFET drain characteristics.

in the pinch-off region for linear amplifiers and in the triode region when used as a variable-resistance element.

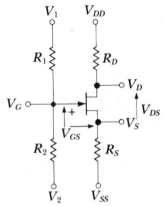

Fig. F-2   CS amplifier circuit.

The analysis can be better understood by defining the following JFET ac and dc parameters:

I_D Drain current.

I_{GSS} Total gate leakage current; usually neglected.

I_{DSS} That value of drain current for V_{DS} equal to or greater than $|V_p|$ and $V_{GS} = 0$ (Fig. F-1).

V_G DC gate voltage when measured with respect to ground (Fig. F-2).

V_D DC drain voltage when measured with respect to ground (Fig. F-2).

V_S DC source voltage when measured with respect to ground (Fig. F-2).

V_{GS} DC gate-to-source voltage, always negative for N-channel and positive for P-channel when used in conventional linear ac small-signal circuits (Fig. F-2).

V_{DS} DC drain-to-source voltage (Fig. F-2).

V_p $V_{GS(off)}$; pinch-off voltage. Two valid definitions: (1) That value of V_{GS} (V_{p1} of Fig. F-1) that causes I_D to be reduced to zero when operated in the pinch-off region; or (2) the magnitude of V_{DS} that causes I_D to level off to a fairly

constant value (V_{p2} of Fig. F-1). Prefix a minus sign for
N-channel and use a plus sign for P-channel when using
V_{p2}.

g_{fs} (y_{fs}) Small-signal transconductance.

g_{fso} (y_{fso}) Value of g_{fs} when $V_{GS}^{*} = 0$.

g_o (y_{os}) Common-source output admittance.

g_{gs} (y_{is}) Common-source input admittance.

r_{DS0} Minimum on resistance that occurs at $V_{GS} = 0$ and $V_{DS} = 0$ in the triode region.

r_{DS} On resistance at any operating point in the triode region.

APPENDIX F1
Pinch-off
Region Analysis

The equation that relates drain current to V_{GS} for the JFET is

$$I_D = I_{DSS} \left(1 - \frac{V_{GS}}{V_p}\right)^2 \tag{F1-1}$$

where the term within brackets is always less than 1. Equation (F1-1) can be manipulated to yield V_{GS} as the dependent variable as follows:

$$V_{GS} = V_p \left[1 - \left(\frac{I_D}{I_{DSS}}\right)^{1/2}\right] \tag{F1-2}$$

The small-signal parameter g_{fso} is related to I_{DSS} and V_p as follows:

$$g_{fso} = \frac{2I_{DSS}}{|V_p|} \tag{F1-3}$$

g_{fs} changes with operating point as follows:

$$g_{fs} = g_{fso}\left(1 - \frac{V_{GS}}{V_p}\right) = \frac{2I_{DSS}}{|V_p|}\left(1 - \frac{V_{GS}}{V_p}\right) \tag{F1-4}$$

$$= g_{fso}\left(\frac{I_D}{I_{DSS}}\right)^{1/2} = \frac{2I_{DSS}}{|V_p|}\left(\frac{I_D}{I_{DSS}}\right)^{1/2} \tag{F1-5}$$

A JFET is biased at its zero temperature coefficient (TC) operating point if

$$V_p - V_{GS} = 0.63 \text{ V} \qquad \text{(F1-6)}$$

Also, at zero drift,

$$I_D = I_{DSS} \left(\frac{0.63}{V_p}\right)^2 \qquad \text{(F1-7)}$$

$$g_{fs} = g_{fso} \left(\frac{0.63}{V_p}\right)^2 \qquad \text{(F1-8)}$$

The gate drift at operating points other than the zero TC point can be calculated as follows:

$$V_G = 2.2\left[1 - \left(\frac{I_D}{I_D \text{ at zero drift}}\right)^{1/2}\right]\frac{\text{mV}}{°\text{C}} \qquad \text{(F1-9)}$$

The analysis of JFETs using Eq. F1-1 can be quite cumbersome since V_{GS} is a function of I_D. However, a technique has been developed that permits the analysis of JFETs without the use of approximations or curves with load lines.* For most all JFET biasing arrangements using resistors and bias voltages, Eq. (F1-1) can be reduced to the general form:

$$I_D = I_{DSS}(\alpha + \beta I_D)^2 \qquad \text{(F1-10)}$$

where α and β are constants determined by the particular biasing circuit. Using the familiar quadratic, I_D can be isolated and solved for:

$$I_D = -\frac{\left(\dfrac{2\alpha}{\beta} - \dfrac{1}{\beta^2 I_{DSS}}\right) \pm \left[\left(\dfrac{2\alpha}{\beta} - \dfrac{1}{\beta^2 I_{DSS}}\right)^2 - \dfrac{4\alpha^2}{\beta^2}\right]^{1/2}}{2} \qquad \text{(F1-11)}$$

Two cursory checks should be made when solving I_D to ensure the answer is a correct one:

    a. Check $V_{GS} = V_G - V_S$ with Eq. (F1-10); V_{GS} should be negative for N-channel and positive for P-channel JFETs.

    b. Check $V_{DS} = V_D - V_S$; this value must be greater in magnitude than the pinch-off voltage V_p or we are in the triode region.

Several common circuit examples have been worked out in Figs. F1-1

---

*Portions of this material taken from Charles R. Hafer, "Bias JFET Circuits Accurately Without Using Curves," *EDN*, March 20, 1978.

through F1-6. The defining equations for these circuits are given on the figures as Eqs. (F1-12) through (F1-43).

An example of a simple circuit will be analyzed to illustrate the ease with which this concept can be used. For the circuit of Fig. F1-1,

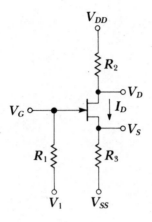

$$V_S = I_D R_3 + V_{SS} \qquad\qquad \text{(F1-12)}$$

$$V_G = V_1 \qquad\qquad \text{(F1-13)}$$

$$V_D = V_{DD} - I_D R_2 \qquad\qquad \text{(F1-14)}$$

$$\alpha = 1 - \frac{V_1 - V_{SS}}{V_p} \qquad\qquad \text{(F1-17)}$$

$$\beta = \frac{R_3}{V_p} \qquad\qquad \text{(F1-18)}$$

(Equations F1-15 and F1-16 are given in the text)

Fig. F1-1   CS N-channel gate bias circuit.

---

assume the following values: $V_1 = -5$ V dc, $V_2 = 0$ V dc, $V_{DD} = 20$ V dc, $V_{SS} = -10$ V dc, $V_p = -5$ V dc, $I_{DSS} = 10$ mA, $R_1 = 100$ kΩ, $R_2 = 5$ kΩ, $R_3 = 5$ kΩ.

The defining equations are:

$$V_S = I_D R_3 + V_{SS}$$

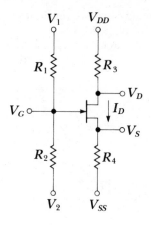

$$\mathbf{V_G} = \frac{\mathbf{R_2(V_1 - V_2)}}{\mathbf{R_1 + R_2}} + \mathbf{V_2} \qquad\qquad (F1\text{-}19)$$

$$V_S = I_D R_4 + V_{SS} \qquad\qquad (F1\text{-}20)$$

$$V_D = V_{DD} - I_D R_3 \qquad\qquad (F1\text{-}21)$$

$$\alpha = 1 - \frac{R_2 V_1}{V_p(R_1 + R_2)} - \frac{R_1 V_2}{V_p(R_1 + R_2)} + \frac{V_{ss}}{V_p} \qquad\qquad (F1\text{-}22)$$

$$\beta = \frac{R_4}{V_p} \qquad\qquad (F1\text{-}23)$$

Fig. F1-2   CS N-channel gate bias circuit.

$$V_G = V_1$$
$$V_D = V_{DD} - I_D R_2$$

Solving for I_D using the relationship $V_{GS} = V_G - V_S$ yields the following:

$$I_D = I_{DSS}\left[1 - \left(\frac{V_1 - I_D R_3 - V_{SS}}{V_p}\right)\right]^2 \qquad (F1\text{-}15)$$

$$I_D = I_{DSS}\left[1 - \left(\frac{V_1 - V_{SS}}{V_p}\right) + \frac{I_D R_3}{V_p}\right]^2 \qquad (F1\text{-}16)$$

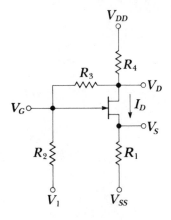

$$V_S = V_{SS} + I_D R_1 \tag{F1-24}$$

$$V_G = V_S + |V_p| \left[\left(\frac{I_D}{I_{DSS}} \right)^{1/2} - 1 \right] \tag{F1-25}$$

$$V_D = V_{DD} - \left(I_D + \frac{V_G - V_1}{R_2} \right) R_4 \tag{F1-26}$$

$$\alpha = 1 + \frac{R_2 V_{DD}}{\lambda |V_p|} + \left[\frac{R_2 R_4 - R_2 \lambda + (R_2 + R_3)\lambda}{\lambda(R_2 + R_3)} \right] \frac{V_1}{|V_p|} - \frac{V_{SS}}{|V_p|} \tag{F1-27}$$

$$\beta = \left(R_1 + \frac{R_2 R_4}{\lambda} \right) \frac{1}{|V_p|} \tag{F1-28}$$

$$\lambda = R_2 + R_3 + R_4$$

Fig. F1-3   *N*-channel voltage feedback bias circuit.

From this expression we obtain

$$\alpha = 1 - \left(\frac{V_1 - V_{SS}}{V_p} \right)$$

$$\beta = R_3/V_p$$

For this circuit, $\alpha = 2$ and $\beta = -1000$.
    Substituting these values into Eq. (F1-11) and solving yields

$$I_D = 1.6 \text{ mA}; \ 2.5 \text{ mA}$$

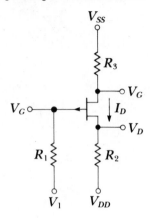

V_{SS} is normally positive.
V_{DD} is normally negative.

$$V_S = V_{SS} - I_D R_3 \qquad\qquad\qquad\text{(F1-29)}$$

$$V_G = V_1 \qquad\qquad\qquad\qquad\text{(F1-30)}$$

$$V_D = V_{DD} + I_D R_2 \qquad\qquad\qquad\text{(F1-31)}$$

$$\alpha = 1 - \frac{V_1 - V_{SS}}{V_p} \qquad\qquad\qquad\text{(F1-32)}$$

$$\beta = \frac{-R_3}{V_p} \qquad\qquad\qquad\qquad\text{(F1-33)}$$

Fig. F1-4   CS P-channel gate bias circuit.

Check $I_D = 2.5$ mA as the solution:

$$V_S = I_D R_3 + V_{SS} = 7.5 + (-10) = -2.5 \text{ V dc}$$
$$V_G = V_1 = -5 \text{ V dc}$$
$$V_{GS} = V_G - V_S = +2.5 \text{ V dc}$$

This is not valid for an N-channel; V_{GS} must be equal to or less than zero. Therefore, $I_D = 2.5$ mA is not an acceptable solution.

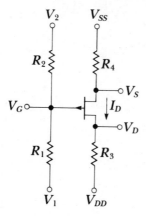

V_{SS} is normally positive.
V_{DD} is normally negative.

$$V_S = V_{SS} - I_D R_4 \qquad \text{(F1-34)}$$

$$V_G = \frac{R_1(V_2 - V_1)}{R_1 + R_2} + V_1 \qquad \text{(F1-35)}$$

$$V_D = V_{DD} + I_D R_3 \qquad \text{(F1-36)}$$

$$\alpha = 1 + \frac{R_1 V_1}{V_p(R_1 + R_2)} - \frac{R_1 V_2}{V_p(R_1 + R_2)} + \frac{V_{SS}}{V_p} - \frac{V_1}{V_p} \qquad \text{(F1-37)}$$

$$\beta = \frac{-R_4}{V_p} \qquad \text{(F1-38)}$$

Fig. F1-5   CS P-channel gate bias circuit.

Check at 1.6 mA:

$$V_S = I_D R_3 + V_{SS} = 8 - 10 = -2 \text{ V dc}$$
$$V_G = V_1 = -5 \text{ V dc}$$
$$V_{GS} = V_G - V_S = -3 \text{ V dc}$$

Check V_{GS} using Eq. (F1-2):

$$V_{GS} = V_p \left[1 - \left(\frac{I_D}{I_{DSS}}\right)^{1/2}\right] = -5\left[1 - \left(\frac{1.6 \times 10^{-3}}{10^{-2}}\right)^{1/2}\right] = -3\text{V dc}$$

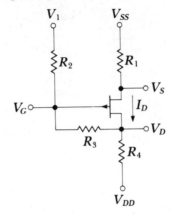

$$V_S = V_{SS} - I_D R_1 \qquad\qquad\qquad\qquad (\text{F1-39})$$

$$V_D = [V_{DD}(R_2 + R_3) + V_1 R_4 + I_D(R_2 + R_3)R_4]\frac{1}{\lambda} \qquad (\text{F1-40})$$

$$V_G = \frac{R_3 V_1}{R_2 + R_3} + \left[\frac{V_{DD}(R_2 + R_3) + V_1 R_4 + I_D(R_2 + R_3)R_4}{\lambda}\right]\frac{R_2}{R_2 + R_3}$$
$$(\text{F1-41})$$

$$\lambda = R_2 + R_3 + R_4$$

$$\alpha = 1 - \frac{1}{V_p}\left[\frac{R_3 V_1}{R_2 + R_3} + \frac{R_2 V_{DD}}{\lambda} + \frac{V_1 R_2 R_4}{\lambda(R_2 + R_3)} - V_{SS}\right] \qquad (\text{F1-42})$$

$$\beta = -\left(\frac{R_2 R_4}{\lambda} + R_1\right)\frac{1}{V_p} \qquad\qquad\qquad (\text{F1-43})$$

Fig. F1-6 P-channel voltage feedback bias circuit.

This checks, and $I_D = 1.6\,\text{mA}$ is an acceptable solution. In general, the smaller value of I_D will yield the correct answer and should be checked first.

APPENDIX F2
JFET Triode
Region Analysis

The triode region is shown in Fig. F-1. The JFET is used in this region because its channel resistance approaches that of a resistor and can be varied as a function of V_{GS}. There are several equations that are useful in the analysis of the JFET in the triode region. The relationship of I_D, V_{GS}, V_{DS}, I_{DSS}, and V_p in the triode region is

$$\frac{I_D}{I_{DSS}} = \frac{2V_{DS}}{V_p} \left(\frac{V_{GS}}{V_p} - \frac{V_{DS}}{2V_p} - 1 \right) \qquad \text{(F2-1)}$$

Define r_{DS0} as the minimum value of r_D and the value of drain resistance when $V_{GS} = 0$; then as V_{GS} varies from 0 V toward pinch-off, we can approximate r_{DS} by

$$r_{DS} = \frac{r_{DS0}}{1 - \left| \dfrac{V_{GS}}{V_p} \right|} \qquad \text{(F2-2)}$$

noting of course that r_{DS} approaches ∞ as V_{GS} approaches V_p. r_{DS0} can be calculated from V_p and I_{DSS} as follows:

$$r_{DS0} = \frac{K|V_p|}{I_{DSS}} = \frac{1}{g_{fs}} \qquad \text{(F2-3)}$$

where $K = 0.4$ to 0.9; normally $K = 0.5$. The channel resistance varies as a function of temperature as follows:

$$r_{DS}(T) = r_{DS}(1 + 0.007\,\Delta T) \qquad \text{(F2-4)}$$

TRANSFER FUNCTION PLOTS FOR TYPICAL TRANSFER FUNCTIONS

A table was compiled by George J. Thaler and Robert G. Brown in their text, "Analysis and Design of Feedback Control Systems," 2d ed., McGraw-Hill, 1960. It provides information relating various nth-order transfer functions with their polar plots, Bode diagrams, Nichols diagrams, and root-locus, respectively. This table is reprinted as a reference and is given in Table G1.

TABLE G-1

$G(s)$	Polar plot	Bode diagram
1. $\dfrac{K}{s\tau_1 + 1}$		
2. $\dfrac{K}{(s\tau_1 + 1)(s\tau_2 + 1)}$		
3. $\dfrac{K}{(s\tau_1+1)(s\tau_2+1)(s\tau_3+1)}$		
4. $\dfrac{K}{s}$		

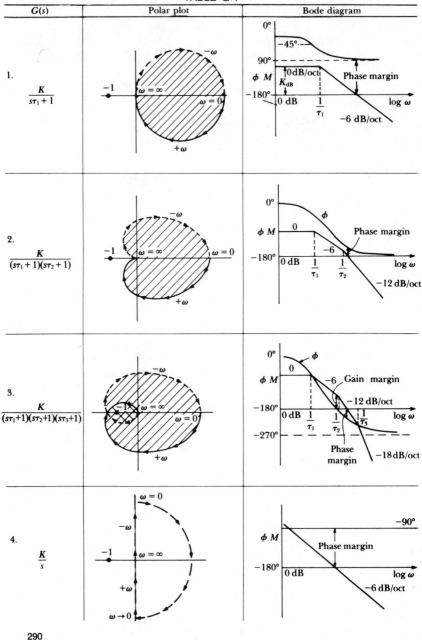

Nichols diagram	Root locus	Comments

Row 1 — Nichols: M, Phase margin, ω, 0 dB, $-180°$ $-90°$ $0°$, ϕ, $\omega = \infty$
Root locus: ω, Root locus, $-\dfrac{1}{\tau_1}$, σ
Comments: Stable; gain margin $= \infty$

Row 2 — Nichols: Phase margin, M, ω, 0 dB, $-180°$ $-90°$ $0°$, ϕ, $\omega \to \infty$
Root locus: ω, R_1, $-\dfrac{1}{\tau_2}$, $-\dfrac{1}{\tau_1}$, σ, R_2
Comments: Elementary regulator; stable; gain margin $= \infty$

Row 3 — Nichols: Phase margin, M, ω, Gain margin, 0 dB, $-180°$ $-90°$ $0°$, ϕ, $\omega \to \infty$
Root locus: ω, R_1, R_3, $-\dfrac{1}{\tau_3}$, $-\dfrac{1}{\tau_2}$, $-\dfrac{1}{\tau_1}$, σ, R_2
Comments: Regulator with additional energy-storage component; unstable, but can be made stable by reducing gain

Row 4 — Nichols: M, Phase margin, ω, 0 dB, $-180°$, $-90°$, ϕ, $\omega \to \infty$
Root locus: ω, σ
Comments: Ideal integrator; stable

$G(s)$	Polar plot	Bode diagram

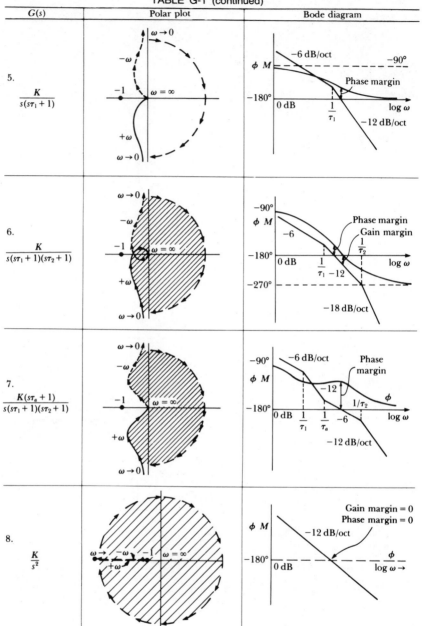

5. $\dfrac{K}{s(s\tau_1 + 1)}$

6. $\dfrac{K}{s(s\tau_1 + 1)(s\tau_2 + 1)}$

7. $\dfrac{K(s\tau_a + 1)}{s(s\tau_1 + 1)(s\tau_2 + 1)}$

8. $\dfrac{K}{s^2}$

Nichols diagram	Root locus	Comments

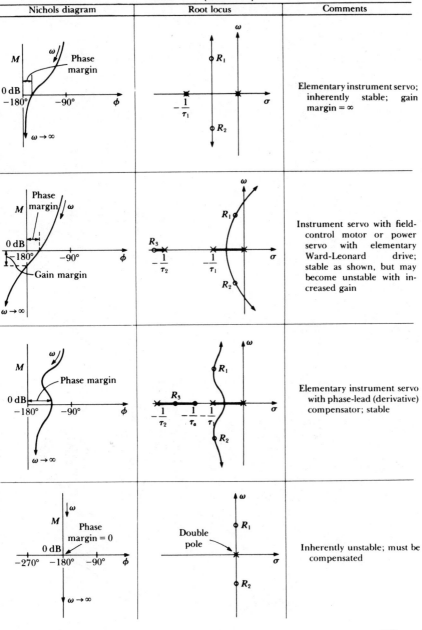

		Elementary instrument servo; inherently stable; gain margin = ∞
		Instrument servo with field-control motor or power servo with elementary Ward-Leonard drive; stable as shown, but may become unstable with increased gain
		Elementary instrument servo with phase-lead (derivative) compensator; stable
		Inherently unstable; must be compensated

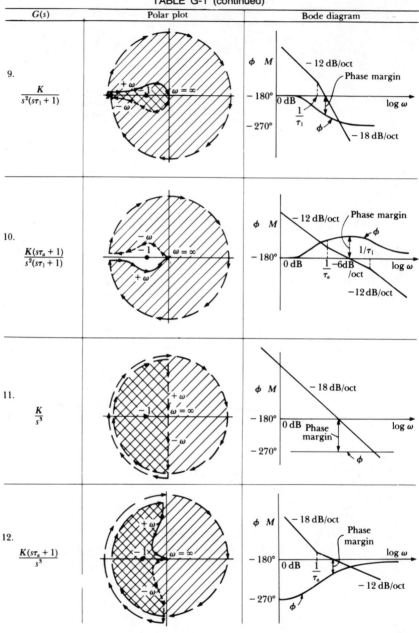

$G(s)$	Polar plot	Bode diagram
9. $\dfrac{K}{s^2(s\tau_1+1)}$		
10. $\dfrac{K(s\tau_a+1)}{s^2(s\tau_1+1)}$		
11. $\dfrac{K}{s^3}$		
12. $\dfrac{K(s\tau_a+1)}{s^3}$		

294

Nichols diagram	Root locus	Comments
M — $0\,dB$, $-270°$, $-180°$, $-90°$, ϕ, Phase margin, $\omega \to \infty$	R_3, R_1 Double pole, σ, ω, R_2	Inherently unstable; must be compensated
M — Phase margin, $0\,dB$, $-180°$, $-90°$, ϕ, $\omega \to \infty$	R_3, R_1 Double pole, $-\frac{1}{\tau_1}$, $-\frac{1}{\tau_a}$, σ, R_2	Stable for all gains
M — Phase margin, $0\,dB$, $-270°$, $-180°$, $-90°$, ϕ, $\omega \to \infty$	R_3, R_1 Triple pole, σ, R_1	Inherently unstable
M — Phase margin, $0\,dB$, $-270°$, $-180°$, $-90°$, ϕ, $\omega \to \infty$	R_3, R_1 Triple pole, $-\frac{1}{\tau_a}$, σ, R_2	Inherently unstable

295

$G(s)$	Polar plot	Bode diagram

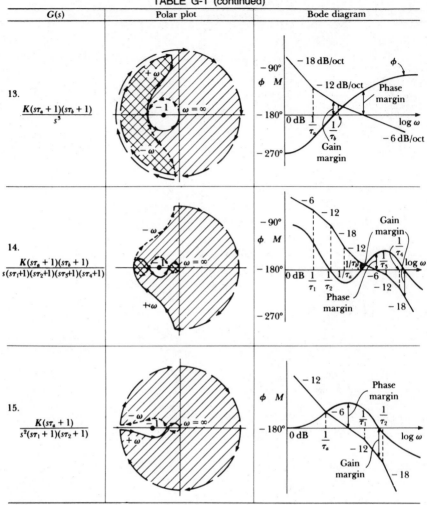

13.

$$\frac{K(s\tau_a + 1)(s\tau_b + 1)}{s^3}$$

14.

$$\frac{K(s\tau_a + 1)(s\tau_b + 1)}{s(s\tau_1+1)(s\tau_2+1)(s\tau_3+1)(s\tau_4+1)}$$

15.

$$\frac{K(s\tau_a + 1)}{s^2(s\tau_1 + 1)(s\tau_2 + 1)}$$

Nichols diagram	Root locus	Comments
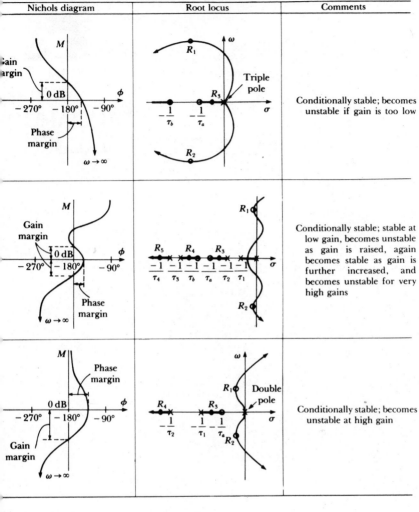		Conditionally stable; becomes unstable if gain is too low
		Conditionally stable; stable at low gain, becomes unstable as gain is raised, again becomes stable as gain is further increased, and becomes unstable for very high gains
		Conditionally stable; becomes unstable at high gain

297

STANDARD
COMPONENT VALUES

APPENDIX H1
Metal-Film Resistor
Standard Values

The multipliers listed in Table H1-1 are applicable for standard metal-film resistors with ±1% initial tolerance. Resistance range is from 10 Ω to 1.5 MΩ.

TABLE H1-1   Metal-film multipliers

1.00	1.47	2.15	3.16	4.64	6.81
1.02	1.50	2.21	3.24	4.75	6.98
1.05	1.54	2.26	3.32	4.87	7.15
1.07	1.58	2.32	3.40	4.99	7.32
1.10	1.62	2.37	3.48	5.11	7.50
1.13	1.65	2.43	3.57	5.23	7.68
1.15	1.69	2.49	3.65	5.36	7.87
1.18	1.74	2.55	3.74	5.49	8.06
1.21	1.78	2.61	3.83	5.62	8.25
1.24	1.82	2.67	3.92	5.76	8.45
1.27	1.87	2.75	4.02	5.90	8.66
1.30	1.91	2.80	4.12	6.04	8.87
1.33	1.96	2.87	4.22	6.19	9.09
1.37	2.00	2.94	4.32	6.34	9.31
1.40	2.05	3.01	4.42	6.49	9.53
1.43	2.10	3.09	4.53	6.65	9.76

APPENDIX H2
Carbon-Composition
Standard Values

The multipliers listed in Table H2-1 are applicable for standard carbon-composition and carbon-film resistors which are ±5% initial tolerances. Typical resistance values range from $0.1\,\Omega$ to $22\,\text{m}\Omega$, depending on type and manufacturer.

TABLE H2-1 Carbon resistor multipliers

1.0	1.6	2.7	4.3	6.8
1.1	1.8	3.0	4.7	7.5
1.2	2.0	3.3	5.1	8.2
1.3	2.2	3.6	5.6	9.1
1.5	2.4	3.9	6.2	

APPENDIX H3
Capacitor Standard Values

In general, the capacitance multipliers listed in Table H3-1 can be obtained in ±5% initial tolerances from 1 pF to 100,000 μF, depending on the capacitor type and manufacturer. For precision capacitors (±1%), the values are too varied to list and depend heavily on the manufacturer.

TABLE H3-1
Capacitance multipliers

1.0	2.7	5.0
1.2	3.0	5.6
1.5	3.3	6.8
2.0	3.9	7.5
2.2	4.3	8.2
2.5	4.7	9.1

FOURIER SERIES
WAVEFORM EQUATIONS

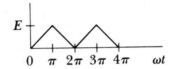

$$e = \frac{E}{2} + \frac{4E \sin \omega t}{\pi^2} - \frac{4E \sin 3\omega t}{9\pi^2} + \frac{4E \sin 5\omega t}{25\pi^2} + \cdots$$

$$+ \frac{4E \sin n\omega t}{n^2\pi^2} \qquad (n \text{ odd})$$

nth term plus if $n = 1, 5, 9, \ldots$

nth term minus if $n = 3, 7, 11, \ldots$ (I-1)

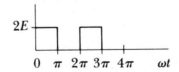

$$e = E + \frac{4E \sin \omega t}{\pi} + \frac{4E \sin 3\omega t}{3\pi} + \frac{4E \sin 5\omega t}{5\pi} + \cdots$$

$$+ \frac{4E \sin n\omega t}{n\pi} \qquad (n \text{ odd}) \qquad\qquad (I-2)$$

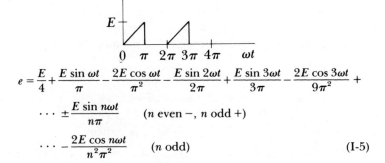

$$e = \frac{2E}{\pi} - \frac{4\cos 2\omega t}{3\pi} - \frac{4E\cos 4\omega t}{15\pi} - \frac{4E\cos 6\omega t}{35\pi} + \cdots$$

$$+ \frac{4E\cos n\omega t}{\pi(1-n^2)} \qquad (n \text{ even}) \tag{I-3}$$

$$e = \frac{E}{\pi} + \frac{E\sin \omega t}{2} - \frac{2E\cos 2\omega t}{3\pi} - \frac{2E\cos 4\omega t}{15\pi} - \frac{2E\cos 6\omega t}{35\pi} + \cdots$$

$$+ \frac{2E\cos n\omega t}{\pi(1-n^2)} \qquad (n \text{ even}) \tag{I-4}$$

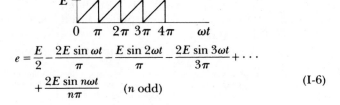

$$e = \frac{E}{4} + \frac{E\sin \omega t}{\pi} - \frac{2E\cos \omega t}{\pi^2} - \frac{E\sin 2\omega t}{2\pi} + \frac{E\sin 3\omega t}{3\pi} - \frac{2E\cos 3\omega t}{9\pi^2} +$$

$$\cdots \pm \frac{E\sin n\omega t}{n\pi} \qquad (n \text{ even} -, n \text{ odd} +)$$

$$\cdots - \frac{2E\cos n\omega t}{n^2\pi^2} \qquad (n \text{ odd}) \tag{I-5}$$

$$e = \frac{E}{2} - \frac{2E\sin \omega t}{\pi} - \frac{E\sin 2\omega t}{\pi} - \frac{2E\sin 3\omega t}{3\pi} + \cdots$$

$$+ \frac{2E\sin n\omega t}{n\pi} \qquad (n \text{ odd}) \tag{I-6}$$

RC NETWORK SYNTHESIS
FOR FEEDBACK AMPLIFIERS

The networks of Table J-1* may be used for the synthesis of various feedback amplifier functions because the transimpedance function $Z_T = v_i/i_o$ is given. The closed-loop gain-bandwidth must be low enough to allow the normal simplifying assumptions to be made. In particular, we will assume the following for the operational amplifier:

$$A_o = \infty \qquad \text{open-loop gain}$$
$$BW = \infty \qquad \text{bandwidth}$$
$$I_B = 0 \qquad \text{bias current}$$
$$R_{in} = \infty \qquad \text{input impedance}$$
$$R_0 = 0 \qquad \text{output impedance}$$

The transimpedance of a circuit is defined as the input voltage divided by the output current when the output is shorted. This is the type transfer function that we need when working with operational amplifiers, because we are summing currents at the amplifier input node. For the circuit of Fig. J-1a, we can derive the transimpedance by referring to Fig. J-1b, where the output is shorted.

*Reproduced by permission of the McGraw-Hill Publishing Company from F. R. Bradley and R. McCoy, "Driftless D-C Amplifier," *Electronics*, April 1952. This table was developed by S. Godet of the Reeves Instrument Corporation, New York.

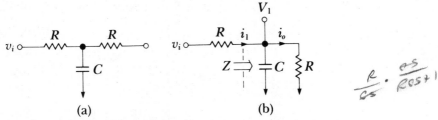

(a)   (b)

Fig. J-1   Circuits for illustrating the transimpedance concept.

$$Z_T = \frac{v_i}{i_o} \qquad\qquad (J\text{-}1)$$

Let
$$Z = R \left\| \frac{1}{Cs} = \frac{(1/Cs)R}{1/Cs + R} = \frac{R}{RCs + 1} \right.$$

$$i_1 = \frac{v_i}{R + Z}$$

$$V_1 = i_1 Z$$

$$i_o = \frac{V_1}{R}$$

Substituting and solving, we find

$$\frac{v_i}{i_o} = 2R(RCs/2 + 1)$$

This is Eq. (J1-3) of Table J-1. The standard configuration usually encountered with an operational amplifier is shown in Fig. J-2.

Note that in the feedback loop the input to the transfer

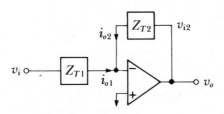

Fig. J-2   Operational amplifier transimpedance circuit.

function is the output of the amplifier. For the amplifier:

$$A_v = \frac{v_o}{v_i} = \frac{v_{i2}}{v_i} \qquad \text{(J-2)}$$

$$v_{i2} = Z_{T2}(i_{o2})$$

$$v_i = Z_{T1}(i_{o1})$$

$$A_v = \frac{Z_{T2}(i_{o2})}{Z_{T1}(i_{o1})}$$

But $\qquad i_{o2} = -i_{o1}$

Therefore

$$A_v = -\frac{Z_{T2}}{Z_{T1}} \qquad \text{(J-3)}$$

EXAMPLE

Let's assume that we want to synthesize a transfer function of

$$A_v = \frac{e_o}{e_i} = \frac{-12}{(s + 0.5)(s + 0.1)}$$

$$= \frac{-12}{(1/2)(2s + 1)(1/10)(10s + 1)} = \frac{-240}{(2s + 1)(10s + 1)}$$

Various combinations could be chosen for Z_{T1} and Z_{T2}, but let's choose:

$$Z_{T1} = 2s + 1$$

$$Z_{T2} = \frac{240}{10s + 1}$$

Referring to Table J-1, Eqs. (J1-2) and (J1-3), we can obtain the circuit shown in Fig. J-3.

In the figure,

$$Z_{T1} = 2s + 1$$

$$R_1 = \frac{A}{2} = \frac{1}{2} \qquad A = 1$$

$$C_1 = \frac{4T}{A} = \frac{8}{1} = 8 \qquad T = 2 \qquad \text{(J1-3)}$$

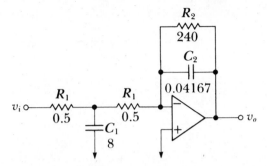

Fig. J-3   Problem example circuit using Table J-1.

From frequency scaling, we can make $R_1 = 500 \, \Omega$ (×1000) and $C_1 = 8000 \, \mu F$ (÷1000).

Similarly,

$$Z_{T2} = \frac{240}{10s + 1}$$

$$R_2 = A = 240 \, \Omega$$

$$C_2 = \frac{T}{A} = \frac{10}{240} = 0.04167 \text{ F} \qquad \text{(J1-2)}$$

Using the same frequency scaling as for Z_{T2}, we can make $R_2 = 240 \text{ k}\Omega$ (×1000) and $C_2 = 41.67 \, \mu F$ (÷ 1000). The circuit becomes that of Fig. J-4.

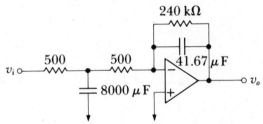

Fig. J-4   Scaled circuit for the example problem.

Let's check: At dc, the gain is

$$A_v = \frac{-240 \times 10^3}{500 + 500} = -240 \qquad \text{(checks)}$$

The transfer function can be checked:

$$i_{o1} = -i_{o2} = \frac{v_i}{Z_{T1}} = \frac{v_i}{A(1 + Ts)} = \frac{v_i}{2R(1 + RCs/2)}$$

$$i_{o2} = \frac{v_i}{(1 \times 10^3)(1 + 2s)}$$

$$v_o = -i_{o2}Z_f$$

$$Z_f = \frac{\left(\dfrac{1}{C_2s}\right)R_2}{R_2 + \dfrac{1}{C_2s}} = \frac{R_2}{R_2C_2s + 1} = \frac{240 \times 10^3}{10s + 1}$$

And solving our standard expression, we find

$$v_o = -Z_f(i_{o2}) = \left(\frac{240 \times 10^3}{10s + 1}\right)\left[\frac{-v_i}{(1 \times 10^3)(1 + 2s)}\right]$$

$$\frac{v_o}{v_i} = \frac{-240}{(1 + 2s)(10s + 1)}$$

$$\frac{v_o}{v_i} = \frac{-12}{\dfrac{(2s + 1)}{2}\dfrac{(10s + 1)}{10}} \qquad \text{(divide by 20)}$$

$$\frac{v_o}{v_i} = \frac{-12}{(s + 0.5)(s + 0.1)} \qquad \text{(this checks)}$$

TABLE J-1

Transfer Impedance Function (v/i)	Network	Relations	Inverse Relations
A (J1-1)		$A = R$	$R = A$
$\dfrac{A}{1+sT}$ (J1-2)		$A = R$ $T = RC$	$R = A$ $C = \dfrac{T}{A}$
$A(1+sT)$ (J1-3)		$A = 2R$ $T = \dfrac{RC}{2}$	$R = \dfrac{A}{2}$ $C = \dfrac{4T}{A}$
$A\left(\dfrac{1+s\theta T}{1+sT}\right)$ $\theta < 1$ (J1-4)	 (a)	$A = R_1 + R_2$ $T = R_2 C$ $\theta = \dfrac{R_1}{R_1 + R_2}$	$R_1 = A\theta$ $R_2 = A(1-\theta)$ $C = \dfrac{T}{A(1-\theta)}$

TABLE J-1 (continued)

Transfer Impedance Function (v_o/i_i)	Network	Relations	Inverse Relations
$A\left(\dfrac{1+s\theta T}{1+sT}\right)$ $\theta < 1$ (J1-4)	(b)	$A = R_1$ $T = (R_1 + R_2)C$ $\theta = \dfrac{R_2}{R_1 + R_2}$	$R_1 = A$ $R_2 = \dfrac{A\theta}{1-\theta}$ $C = \dfrac{T(1-\theta)}{A}$
$A\left(\dfrac{1+sT}{1+s\theta T}\right)$ $\theta < 1$	(a)	$A = \dfrac{2R_1 R_2}{2R_1 + R_2}$ $T = \dfrac{R_2 C}{2}$ $\theta = \dfrac{2R_1}{2R_1 + R_2}$	$R_1 = \dfrac{A}{2(1-\theta)}$ $R_2 = \dfrac{A}{\theta}$ $C = \dfrac{4T(1-\theta)}{A}$
	(b)	$A = 2R_1$ $T = \left(R_2 + \dfrac{R_1}{2}\right)C$ $\theta = \dfrac{2R_2}{2R_2 + R_1}$	$R_1 = \dfrac{A}{2}$ $R_2 = \dfrac{A\theta}{4(1-\theta)}$ $C = \dfrac{4T(1-\theta)}{A}$

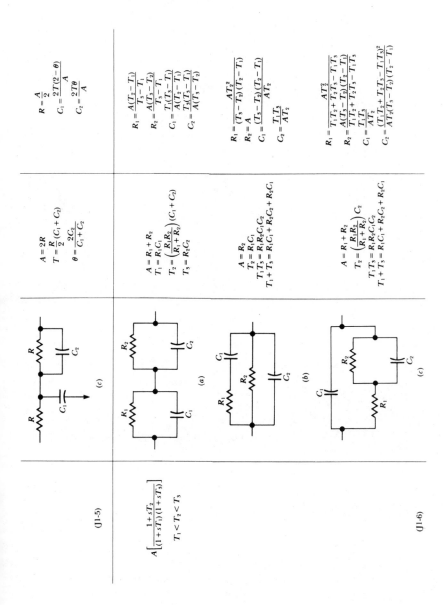

(J1-5)

$$A\left[\frac{1+sT_2}{(1+sT_1)(1+sT_3)}\right]$$

$$T_1 < T_2 < T_3$$

(J1-6)

(c)

$$R = \frac{A}{2}$$
$$C_1 = \frac{2T(2-\theta)}{A}$$
$$C_2 = \frac{2T\theta}{A}$$

$$A = 2R$$
$$T = \frac{R}{2}(C_1+C_2)$$
$$\theta = \frac{2C_2}{C_1+C_2}$$

(a)

$$R_1 = \frac{A(T_2-T_1)}{T_3-T_1}$$
$$R_2 = \frac{A(T_3-T_2)}{T_3-T_1}$$
$$C_1 = \frac{T_1(T_3-T_1)}{A(T_2-T_1)}$$
$$C_2 = \frac{T_3(T_3-T_1)}{A(T_3-T_2)}$$

$$A = R_1+R_2$$
$$T_1 = R_1 C_1$$
$$T_2 = \left(\frac{R_1 R_2}{R_1+R_2}\right)(C_1+C_2)$$
$$T_3 = R_2 C_2$$

(b)

$$R_1 = \frac{AT_2^2}{(T_3-T_2)(T_2-T_1)}$$
$$R_2 = A$$
$$C_1 = \frac{(T_3-T_2)(T_2-T_1)}{AT_2}$$
$$C_2 = \frac{T_1 T_3}{AT_2}$$

$$A = R_2$$
$$T_2 = R_2 C_1$$
$$T_1 T_3 = R_1 R_2 C_1 C_2$$
$$T_1+T_3 = R_1 C_1 + R_2 C_2 + R_2 C_1$$

(c)

$$R_1 = \frac{AT_2^2}{T_1 T_2 + T_2 T_3 - T_1 T_3}$$
$$R_2 = \frac{A(T_3-T_2)(T_2-T_1)}{T_1 T_2 + T_2 T_3 - T_1 T_3}$$
$$C_1 = \frac{T_1 T_3}{AT_2}$$
$$C_2 = \frac{(T_1 T_2 + T_2 T_3 - T_1 T_3)^2}{AT_2(T_3-T_2)(T_2-T_1)}$$

$$A = R_1+R_2$$
$$T_2 = \left(\frac{R_1 R_2}{R_1+R_2}\right)C_2$$
$$T_1 T_3 = R_1 R_2 C_1 C_2$$
$$T_1+T_3 = R_1 C_1 + R_2 C_2 + R_2 C_1$$

TABLE J-1 (continued)

Transfer Impedance Function (v_i/i_i)	Network	Relations	Inverse Relations
$A\left[\dfrac{1+sT_2}{(1+sT_1)(1+sT_3)}\right]$ $T_1 < T_2 < T_3$ (J1-6)	(d)	$A = R_1$ $T_2 = R_2(C_1 + C_2)$ $T_1 T_3 = R_1 R_2 C_1 C_2$ $T_1 + T_3 = R_1 C_1 + R_2 C_2 + R_2 C_1$	$R_1 = A$ $R_2 = \dfrac{A(T_3 - T_2)(T_2 - T_1)}{(T_1 + T_3 - T_2)^2}$ $C_1 = \dfrac{T_1 + T_3 - T_2}{A}$ $C_2 = \dfrac{T_1 T_3(T_1 + T_3 - T_2)}{A(T_3 - T_2)(T_2 - T_1)}$
$A\left[\dfrac{1+sT_2}{(1+sT_1)(1+sT_3)}\right]$ $T_2 \ll T_1 \ll T_3$ (J1-7)		$A = 2R_1 + \dfrac{R_1^2}{R_2}$ $T_1 = R_1 C_1$ $T_2 = \left(\dfrac{R_1 R_2}{R_1 + 2R_2}\right)(C_1 + C_2)$ $T_3 = R_1 C_2$	$R_1 = \dfrac{AT_2}{(T_1 + T_3)}$ $R_2 = \dfrac{(T_1 + T_3)(T_1 + T_3 - 2T_2)}{AT_2^2}$ $C_1 = \dfrac{T_1(T_1 + T_3)}{AT_2}$ $C_2 = \dfrac{T_3(T_1 + T_3)}{AT_2}$
$A\left[\dfrac{1+sT_2}{(1+sT_1)(1+sT_3)}\right]$ $T_1 \ll T_3 \ll T_2$ (J1-8)		$A = R_1 + R_2$ $T_1 = R_1 C_1$ $T_2 = \dfrac{R_1 R_2}{R_1 + R_2}(2C_1 + C_2)$ $T_3 = R_2 C_1$	$R_1 = \dfrac{AT_1}{(T_1 + T_3)}$ $R_2 = \dfrac{AT_3}{(T_1 + T_3)}$ $C_1 = \dfrac{A}{(T_1 + T_3)}$ $C_2 = \dfrac{(T_1 + T_3)}{A}\left(\dfrac{T_2}{T_3} + \dfrac{T_2}{T_1} - 2\right)$

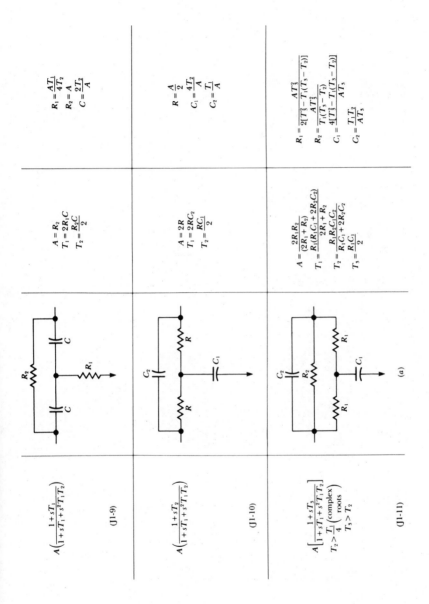

$$A\left(\frac{1+sT_1}{1+sT_1+s^2T_1T_2}\right)$$

(J1-9)

$A = R_2$
$T_1 = 2R_1C$
$T_2 = \dfrac{R_2C}{2}$

$R_1 = \dfrac{AT_1}{4T_2}$
$R_2 = A$
$C = \dfrac{2T_2}{A}$

$$A\left(\frac{1+sT_2}{1+sT_1+s^2T_1T_2}\right)$$

(J1-10)

$A = 2R$
$T_1 = 2RC_2$
$T_2 = \dfrac{RC_1}{2}$

$R = \dfrac{A}{2}$
$C_1 = \dfrac{4T_2}{A}$
$C_2 = \dfrac{T_1}{A}$

$$A\left[\frac{1+sT_3}{1+sT_1+s^2T_1T_2}\right]$$
$T_2 > \dfrac{T_1}{4}\left(\begin{array}{c}\text{complex}\\\text{roots}\end{array}\right)$
$T_3 > T_2$

(J1-11)

$A = \dfrac{2R_1R_2}{(2R_1+R_2)}$
$T_1 = \dfrac{R_1(R_1C_1+2R_2C_2)}{2R_1+R_2}$
$T_2 = \dfrac{R_2C_1C_2}{R_1C_1+2R_2C_2}$
$T_3 = \dfrac{R_1C_1}{2}$

$R_1 = \dfrac{AT_3^2}{2[T_2^2-T_1(T_3-T_2)]}$
$R_2 = \dfrac{AT_3^2}{T_1(T_3-T_2)}$
$C_1 = \dfrac{4[T_3^2-T_1(T_3-T_2)]}{AT_3}$
$C_2 = \dfrac{T_1T_2}{AT_3}$

(a)

311

TABLE J-1 (continued)

Transfer Impedance Function (v/i_o)	Network	Relations	Inverse Relations
$A\left[\dfrac{1+sT_3}{1+sT_1+s^2T_1T_2}\right]$ $T_2 > \dfrac{T_1}{4}$ (complex roots) $T_3 > T_2$	(b)	$A = 2R_1$ $T_1 = R_2C_1 + 2R_1C_2$ $T_2 = \dfrac{R_1(R_1 + 2R_2)C_1C_2}{R_2C_1 + 2R_1C_2}$ $T_3 = \left(R_2 + \dfrac{R_1}{2}\right)C_1$	$R_1 = \dfrac{A}{2}$ $R_2 = \dfrac{AT_1(T_3 - T_2)}{4[T_3^2 - T_1(T_3 - T_2)]}$ $C_1 = \dfrac{4[T_3^2 - T_1(T_3 - T_2)]}{AT_3}$ $C_2 = \dfrac{T_1T_2}{AT_3}$
	(c)	$A = 2R$ $T_1 = R(C_2 + 2C_3)$ $T_2 = \dfrac{RC_3(C_1 + C_2)}{C_2 + 2C_3}$ $T_3 = \dfrac{R}{2}(C_1 + C_2)$	$R = \dfrac{A}{2}$ $C_1 = \dfrac{2[2T_3 - T_1(T_3 - T_2)]}{AT_3}$ $C_2 = \dfrac{2T_1(T_3 - T_2)}{AT_3}$ $C_3 = \dfrac{T_1T_2}{AT_3}$

(J1-11)

$$A\left[\frac{1+sT_3}{1+sT_1+s^2T_1T_2}\right]$$

$$T_2 > \frac{T_1}{4} \text{ (complex roots)}$$

$$T_3 < T_1$$

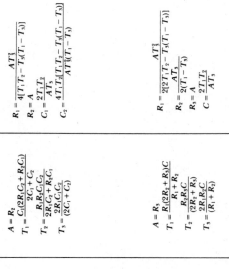

(a)

$$A = R_2$$
$$T_1 = 2R_1C_1 + R_2C_2$$
$$T_2 = \frac{R_1R_2C_1(C_1+2C_2)}{2R_1C_1+R_2C_2}$$
$$T_3 = 2R_1C_1$$

$$R_1 = \frac{AT_3^2}{4[T_1T_2 - T_3(T_1-T_3)]}$$
$$R_2 = A$$
$$C_1 = \frac{2[T_1T_2 - T_3(T_1-T_3)]}{AT_3}$$
$$C_2 = \frac{(T_1-T_3)}{A}$$

(b)

$$A = R_2$$
$$T_1 = \frac{C_1(2R_1C_2 + R_2C_1)}{2C_1+C_2}$$
$$T_2 = \frac{R_1R_2C_1C_2}{2R_1C_2+R_2C_1}$$
$$T_3 = \frac{2R_1C_1C_2}{(2C_1+C_2)}$$

$$R_1 = \frac{AT_3^2}{4[T_1T_2 - T_3(T_1-T_3)]}$$
$$R_2 = A$$
$$C_1 = \frac{2T_1T_2}{AT_3}$$
$$C_2 = \frac{4T_1T_2[T_1T_2 - T_3(T_1-T_3)]}{AT_3^2(T_1-T_3)}$$

(c)

$$A = R_3$$
$$T_1 = \frac{R_1(2R_2+R_3)C}{R_1+R_2}$$
$$T_2 = \frac{R_2R_3C}{2R_2+R_3}$$
$$T_3 = \frac{2R_1R_2C}{(R_1+R_2)}$$

$$R_1 = \frac{AT_3^2}{2[2T_1T_2 - T_3(T_1-T_3)]}$$
$$R_2 = \frac{2(T_1-T_3)}{AT_3}$$
$$R_3 = A$$
$$C = \frac{2T_1T_2}{AT_3}$$

(J1-12)

313

TABLE J-1 (continued)

Transfer Impedance Function (v/i_k)	Network	Relations	Inverse Relations
$A(1+sT_1)(1+sT_2)$ $T_1 < T_2$ (J1-13)		$A = 2R_1 + R_2$ $T_1 = \left(\dfrac{R_1 R_2}{2R_1 + R_2}\right)C$ $T_2 = R_1 C$	$R_1 = A\left(\dfrac{T_2 - T_1}{2T_2}\right)$ $R_2 = A\dfrac{T_1}{T_2}$ $C = \dfrac{2T_2}{A(T_2 - T_1)}$
$A\left(\dfrac{1+sT_1}{1+s^2 T_1 T_2}\right)$ (J1-14)	$R_1 C_1 = 4R_2 C_2$	$A = 2R_1$ $T_1 = \dfrac{R_1 C_1}{2} = 2R_2 C_2$ $T_2 = R_1 C_2$	$R_1 = \dfrac{A}{2}$ $R_2 = \dfrac{AT_1}{4T_2}$ $C_1 = \dfrac{4T_1}{A}$ $C_2 = \dfrac{2T_2}{A}$
$\dfrac{1}{sB}$ (J1-15)		$B = C$	$C = B$
$\dfrac{1}{sB}(1 + sT_3)$ (J1-16)		$B = C$ $T = RC$	$R = \dfrac{T}{B}$ $C = B$

Transfer function	Circuit		
$\dfrac{1}{B}\left(\dfrac{1+sT}{s^2T}\right)$ (J1-17)	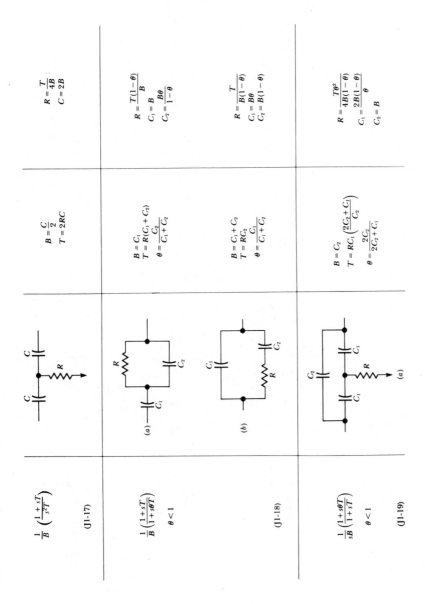	$B=\dfrac{C}{2}$ $T=2RC$	$R=\dfrac{T}{4B}$ $C=2B$
$\dfrac{1}{B}\left(\dfrac{1+sT}{1+s\theta T}\right)$ $\theta<1$ (J1-18)	*(a)*	$B=C_1$ $T=R(C_1+C_2)$ $\theta=\dfrac{C_2}{C_1+C_2}$	$R=\dfrac{T(1-\theta)}{B}$ $C_1=B$ $C_2=\dfrac{B\theta}{1-\theta}$
	(b)	$B=C_1+C_2$ $T=RC_2$ $\theta=\dfrac{C_1}{C_1+C_2}$	$R=\dfrac{T}{B(1-\theta)}$ $C_1=B\theta$ $C_2=B(1-\theta)$
$\dfrac{1}{sB}\left(\dfrac{1+s\theta T}{1+sT}\right)$ $\theta<1$ (J1-19)	*(a)*	$B=C_2$ $T=RC_1\left(\dfrac{2C_2+C_1}{C_2}\right)$ $\theta=\dfrac{2C_2}{2C_2+C_1}$	$R=\dfrac{T\theta^2}{4B(1-\theta)}$ $C_1=\dfrac{2B(1-\theta)}{\theta}$ $C_2=B$

315

TABLE J-1 (continued)

Transfer Impedance Function (v_o/i_o)	Network	Relations	Inverse Relations
$\dfrac{1}{sB}\left(\dfrac{1+s\theta T}{1+sT}\right)$ $\theta < 1$	(b)	$B=\dfrac{C_1^2}{2C_1+C_2}$ $T=RC_2$ $\theta=\dfrac{2C_1}{2C_1+C_2}$	$R=\dfrac{T\theta^2}{4B(1-\theta)}$ $C_1=\dfrac{2B}{\theta}$ $C_2=\dfrac{4B(1-\theta)}{\theta^2}$
(J1-19)	(c)	$B=\left(\dfrac{R_1}{R_1+R_2}\right)C$ $T=R_2C$ $\theta=\dfrac{2R_1}{R_1+R_2}$	$R_1=\dfrac{T\theta^2}{2B(2-\theta)}$ $R_2=\dfrac{T\theta}{2B}$ $C=\dfrac{2B}{\theta}$
$\dfrac{1}{sB}\left[\dfrac{(1+sT_1)(1+sT_3)}{1+sT_2}\right]$ $T_1 < T_2 < T_3$	(a)	$B=C_1+C_2$ $T_1=R_1C_1$ $T_2=(R_1+R_2)\left(\dfrac{C_1C_2}{C_1+C_2}\right)$ $T_3=R_2C_2$	$R_1=\dfrac{T_1(T_3-T_1)}{B(T_2-T_1)}$ $R_2=\dfrac{T_3(T_3-T_2)}{B(T_3-T_1)}$ $C_1=\dfrac{T_3-T_1}{T_2-T_1}$ $C_2=\dfrac{B(T_3-T_2)}{T_3-T_1}$

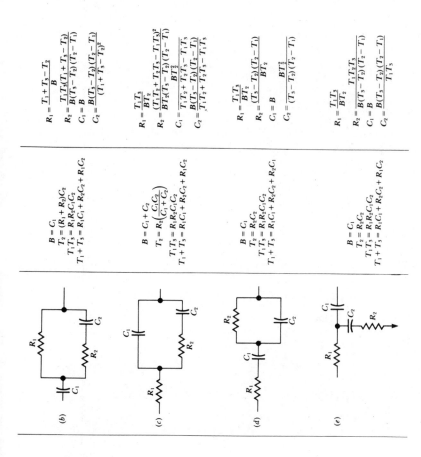

Circuit		
(b)	$B = C_1$ $T_2 = (R_1 + R_2)C_2$ $T_1 T_3 = R_1 R_2 C_1 C_2$ $T_1 + T_3 = R_1 C_1 + R_2 C_2 + R_1 C_2$	$R_1 = \dfrac{T_1 + T_3 - T_2}{B}$ $R_2 = \dfrac{T_1 T_3 (T_1 + T_3 - T_2)}{B(T_3 - T_2)(T_2 - T_1)}$ $C_1 = B$ $C_2 = \dfrac{B(T_3 - T_2)(T_2 - T_1)}{(T_1 + T_3 - T_2)^2}$
(c)	$B = C_1 + C_2$ $T_2 = R_2\left(\dfrac{C_1 C_2}{C_1 + C_2}\right)$ $T_1 T_3 = R_1 R_2 C_1 C_2$ $T_1 + T_3 = R_1 C_1 + R_2 C_2 + R_1 C_2$	$R_1 = \dfrac{T_1 T_3}{B T_2^2}$ $R_2 = \dfrac{(T_1 T_2 + T_2 T_3 - T_1 T_3)^2}{B T_2 (T_3 - T_2)(T_2 - T_1)}$ $C_1 = \dfrac{T_1 T_2 + T_2 T_3 - T_1 T_3}{B T_2^2}$ $C_2 = \dfrac{B(T_3 - T_2)(T_2 - T_1)}{T_1 T_2 + T_2 T_3 - T_1 T_3}$
(d)	$B = C_1$ $T_2 = R_2 C_2$ $T_1 T_3 = R_1 R_2 C_1 C_2$ $T_1 + T_3 = R_1 C_1 + R_2 C_2 + R_2 C_1$	$R_1 = \dfrac{T_1 T_3}{B T_2}$ $R_2 = \dfrac{(T_3 - T_2)(T_2 - T_1)}{B T_2}$ $C_1 = B$ $C_2 = \dfrac{B T_2^2}{(T_3 - T_2)(T_2 - T_1)}$
(e)	$B = C_1$ $T_2 = R_2 C_2$ $T_1 T_3 = R_1 R_2 C_1 C_2$ $T_1 + T_3 = R_1 C_1 + R_2 C_2 + R_1 C_2$	$R_1 = \dfrac{T_1 T_3}{B T_2}$ $R_2 = \dfrac{B(T_3 - T_2)(T_2 - T_1)}{T_1 T_2 T_3}$ $C_1 = B$ $C_2 = \dfrac{B(T_3 - T_2)(T_2 - T_1)}{T_1 T_3}$

(J1-20)

TABLE J-1 (continued)

Transfer Impedance Function (v_0/i_0)	Network	Relations	Inverse Relations
$\dfrac{1}{sB}(1+sT_1)(1+sT_2)$ $T_1 \neq T_2$ (J1-21)		$B = C_2$ $T_1 T_2 = R_1 R_2 C_1 C_2$ $T_1 + T_2 = R_1 C_1 + R_2 C_2 + R_1 C_2$	$R_1 = \dfrac{[(T_1)^{1/2} - (T_2)^{1/2}]^2}{B}$ $R_2 = \dfrac{(T_1 T_2)^{1/2}}{B}$ $C_1 = \dfrac{B(T_1 T_2)^{1/2}}{[(T_1)^{1/2} - (T_2)^{1/2}]^2}$ $C_2 = B$
$\dfrac{1}{sB}\left[\dfrac{(1+sT_1)(1+sT_2)}{s\sqrt{T_1 T_2}}\right]$ $T_1 \neq T_2$ (J1-22)		$B = C_2$ $T_1 T_2 = R_1 R_2 C_1 C_2$ $T_1 + T_2 = R_1 C_1 + R_2 C_2 + R_1 C_2$	$R_1 = \dfrac{[(T_1)^{1/2} - (T_2)^{1/2}]^2}{B}$ $R_2 = \dfrac{(T_1 T_2)^{1/2}}{B}$ $C_1 = \dfrac{B(T_1 T_2)^{1/2}}{[(T_1)^{1/2} - (T_2)^{1/2}]^2}$ $C_2 = B$
$\dfrac{1}{sB}\left[\dfrac{(1+sT_1)(1+sT_2)}{s^2 T_1 T_2}\right]$ $T_1 < T_2$ (J1-23)		$B = \dfrac{C_1 C_2}{C_1 + 2C_2}$ $T_1 = RC_1$ $T_2 = R(C_1 + 2C_2)$	$R = \dfrac{T_1(T_2 - T_1)}{2BT_2}$ $C_1 = \dfrac{2BT_2}{T_2 - T_1}$ $C_2 = \dfrac{BT_2}{T_1}$

RC TRANSFER
FUNCTION EQUATIONS

The information contained in this appendix is useful in solving problems where transfer functions are involved: $G(s) = e_o(s)/e_{in}(s)$. Each transfer function is represented by one or more Bode plots; and each Bode plot is represented by one possible circuit realization. Table K-1 provides the relationships among the transfer function, Bode plot, and circuit configuration and defines the relations between time constants and circuit element values.

TABLE K-1

Transfer Function	Bode Plot	Circuit	Relations
$\dfrac{1}{Ts+1}$ (K1-1)			$T = R_1 C_1$
$\dfrac{A}{Ts+1}$ (K1-2)			$A = \dfrac{R_2}{R_1 + R_2}$ $T = \dfrac{R_1 R_2 C_1}{R_1 + R_2}$
$\dfrac{Ts}{Ts+1}$ (K1-3)			$T = R_1 C_1$

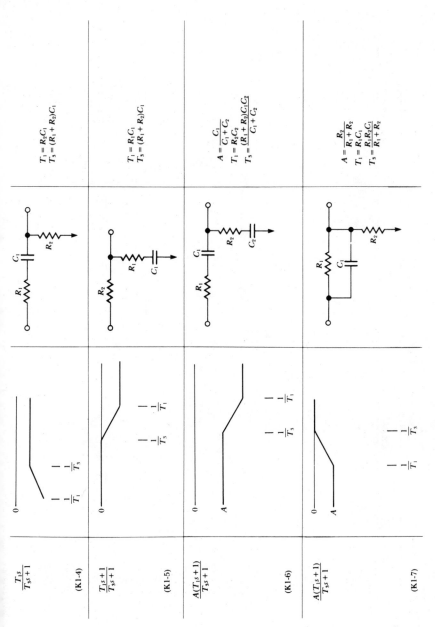

$$\frac{T_1 s}{T_3 s + 1}$$

(K1-4)

$$T_1 = R_2 C_1$$
$$T_3 = (R_1 + R_2)C_1$$

$$\frac{T_1 s + 1}{T_3 s + 1}$$

(K1-5)

$$T_1 = R_1 C_1$$
$$T_3 = (R_1 + R_2)C_1$$

$$\frac{A(T_1 s + 1)}{T_3 s + 1}$$

(K1-6)

$$A = \frac{C_1}{C_1 + C_2}$$
$$T_1 = R_2 C_2$$
$$T_3 = \frac{(R_1 + R_2)C_1 C_2}{C_1 + C_2}$$

$$\frac{A(T_1 s + 1)}{T_3 s + 1}$$

(K1-7)

$$A = \frac{R_2}{R_1 + R_2}$$
$$T_1 = R_1 C_1$$
$$T_3 = \frac{R_1 R_2 C_1}{R_1 + R_2}$$

321

TABLE K-1 (continued)

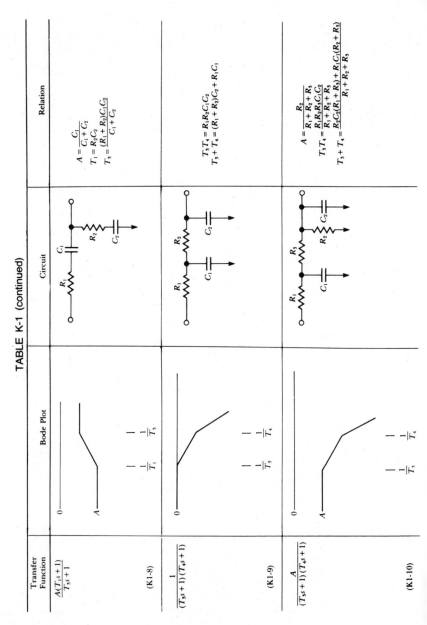

Transfer Function	Bode Plot	Circuit	Relation
$\dfrac{A(T_1s+1)}{T_3s+1}$ (K1-8)			$A = \dfrac{C_1}{C_1+C_2}$ $T_1 = R_2C_2$ $T_3 = \dfrac{(R_1+R_2)C_1C_2}{C_1+C_2}$
$\dfrac{1}{(T_3s+1)(T_4s+1)}$ (K1-9)			$T_3T_4 = R_2C_1C_2$ $T_3+T_4 = (R_1+R_2)C_2 + R_1C_1$
$\dfrac{A}{(T_3s+1)(T_4s+1)}$ (K1-10)			$A = \dfrac{R_2}{R_1+R_2+R_3}$ $T_3T_4 = \dfrac{R_2R_3C_1C_2}{R_1+R_2+R_3}$ $T_3+T_4 = \dfrac{R_2C_2(R_1+R_3)+R_1C_1(R_2+R_3)}{R_1+R_2+R_3}$

322

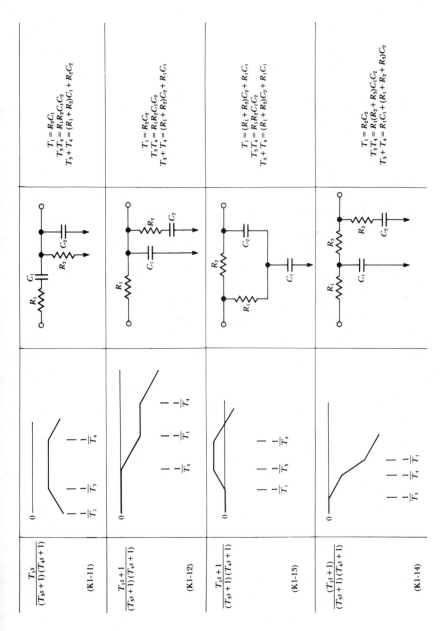

$$\frac{T_1 s}{(T_3 s+1)(T_4 s+1)}$$

(K1-11)

$T_1 = R_2 C_1$
$T_3 T_4 = R_2 R_3 C_1 C_2$
$T_3 + T_4 = (R_1 + R_2)C_1 + R_2 C_2$

$$\frac{T_1 s+1}{(T_3 s+1)(T_4 s+1)}$$

(K1-12)

$T_1 = R_2 C_2$
$T_3 T_4 = R_1 R_2 C_1 C_2$
$T_3 + T_4 = (R_1 + R_2)C_2 + R_1 C_1$

$$\frac{T_1 s+1}{(T_3 s+1)(T_4 s+1)}$$

(K1-13)

$T_1 = (R_1 + R_2)C_2 + R_1 C_1$
$T_3 T_4 = R_1 R_2 C_1 C_2$
$T_3 + T_4 = (R_1 + R_2)C_2 + R_1 C_1$

$$\frac{(T_1 s+1)}{(T_3 s+1)(T_4 s+1)}$$

(K1-14)

$T_1 = R_2 C_2$
$T_3 T_4 = R_1(R_2 + R_3)C_1 C_2$
$T_3 + T_4 = R_3 C_1 + (R_1 + R_2 + R_3)C_2$

323

TABLE K-1 (continued)

Transfer Function	Bode Plot	Circuit	Relation
$\dfrac{A(T_1s+1)}{(T_3s+1)(T_4s+1)}$ (K1-15)			$A = \dfrac{R_2}{R_2+R_3}$ $T_1 = (R_1+R_3)C_1$ $T_3T_4 = \dfrac{R_1R_2R_3C_1C_2}{R_2+R_3}$ $T_3+T_4 = R_1C_1 + \dfrac{R_2R_3(C_1+C_2)}{R_2+R_3}$
$\dfrac{A(T_1s+1)}{(T_3s+1)(T_4s+1)}$ (K1-16)			$A = \dfrac{R_3}{R_1+R_3}$ $T_1 = (R_1+R_2)C_1 + R_2C_2$ $T_3T_4 = \dfrac{R_1R_2R_3C_1C_2}{R_1+R_3}$ $T_3+T_4 = R_2(C_1+C_2) + \dfrac{R_1R_3C_1}{R_1+R_3}$
$\dfrac{T_1T_2s^2}{(T_3s+1)(T_4s+1)}$ (K1-17)			$T_1T_2 = T_3T_4 = R_1R_2C_1C_2$ $T_1+T_2 = R_1(C_1+C_2) + R_2C_2$ $T_3+T_4 = T_1+T_2$

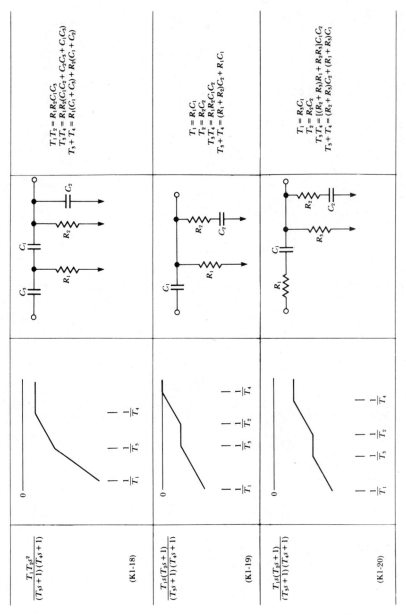

$$\frac{T_1 T_2 s^2}{(T_3 s+1)(T_4 s+1)}$$

(K1-18)

$T_1 T_2 = R_1 R_2 C_1 C_3$
$T_3 T_4 = R_1 R_2 (C_1 C_2 + C_2 C_3 + C_1 C_3)$
$T_3 + T_4 = R_1(C_1 + C_3) + R_2(C_1 + C_2)$

$$\frac{T_1 s(T_2 s+1)}{(T_3 s+1)(T_4 s+1)}$$

(K1-19)

$T_1 = R_1 C_1$
$T_2 = R_2 C_2$
$T_3 T_4 = R_1 R_2 C_1 C_2$
$T_3 + T_4 = (R_1 + R_2)C_2 + R_1 C_1$

$$\frac{T_1 s(T_2 s+1)}{(T_3 s+1)(T_4 s+1)}$$

(K1-20)

$T_1 = R_3 C_1$
$T_2 = R_2 C_2$
$T_3 T_4 = [(R_2 + R_3)R_1 + R_2 R_3]C_1 C_2$
$T_3 + T_4 = (R_2 + R_3)C_2 + (R_1 + R_3)C_1$

TABLE K-1 (continued)

Transfer Function	Bode Plot	Circuit	Relation
$\dfrac{T_1 s(T_2 s + 1)}{(T_3 s + 1)(T_4 s + 1)}$ (K1-21)	markings: $\dfrac{1}{T_1}$, $\dfrac{1}{T_3}$, $\dfrac{1}{T_2}$, $\dfrac{1}{T_4}$	(circuit with C_2, R_2, C_1, R_1)	$T_1 = (R_1 + R_2)C_2 + R_1 C_1$ $T_2 = \dfrac{R_1(C_1 + C_2) + R_2 C_2}{R_1 R_2 C_1 C_2}$ $T_3 T_4 = R_1 R_2 C_1 C_2$ $T_3 + T_4 = (R_1 + R_2)C_2 + R_1 C_1$
$\dfrac{T_1 s(T_2 s + 1)}{(T_3 s + 1)(T_4 s + 1)}$ (K1-22)	markings: $\dfrac{1}{T_1}$, $\dfrac{1}{T_3}$, $\dfrac{1}{T_4}$, $\dfrac{1}{T_2}$	(circuit with C_2, R_2, R_1, C_1, R_3)	$T_1 = R_3 C_1 + (R_2 + R_3)C_2$ $T_2 = \dfrac{R_2 R_3 C_1 C_2}{R_3 C_1 + (R_2 + R_3)C_2}$ $T_3 T_4 = R_2(R_1 + R_3)C_1 C_2$ $T_3 + T_4 = (R_1 + R_3)(C_1 + C_2) + R_2 C_2$
$\dfrac{T_1 s(T_2 s + 1)}{(T_3 s + 1)(T_4 s + 1)}$ (K1-23)	markings: $\dfrac{1}{T_1}$, $\dfrac{1}{T_2}$, $\dfrac{1}{T_3}$, $\dfrac{1}{T_4}$	(circuit with C_1, R_2, C_2, R_3, R_1)	$T_1 = \dfrac{R_1 R_3 C_1}{R_1 + R_2 + R_3}$ $T_2 = R_2 C_2$ $T_3 T_4 = \dfrac{R_1 R_2 R_3 C_1 C_2}{R_1 + R_2 + R_3}$ $T_3 + T_4 = \dfrac{R_2 C_2(R_1 + R_3) + R_1 C_1(R_2 + R_3)}{R_1 + R_2 + R_3}$

Transfer function	Circuit	Bode plot	Time constants
$$\frac{(T_1 s + 1)(T_2 s + 1)}{(T_3 s + 1)(T_4 s + 1)}$$ (K1-24)		$-\frac{1}{T_3} \quad -\frac{1}{T_1} \quad -\frac{1}{T_2} \quad -\frac{1}{T_4}$	$T_1 = R_1 C_1$ $T_2 = R_2 C_2$ $T_3 T_4 = R_1 R_2 C_1 C_2$ $T_3 + T_4 = (R_1 + R_2) C_2 + R_1 C_1$
$$\frac{(T_1 s + 1)(T_2 s + 1)}{(T_3 s + 1)(T_4 s + 1)}$$ (K1-25)		$-\frac{1}{T_3} \quad -\frac{1}{T_1} \quad -\frac{1}{T_4} \quad -\frac{1}{T_2}$	$T_1 = R_2 C_2$ $T_2 = (R_1 + R_3) C_1$ $T_3 T_4 = [(R_1 + R_3) R_2 + R_1 R_3] C_1 C_2$ $T_3 + T_4 = (R_1 + R_3) C_1 + (R_2 + R_3) C_2$
$$\frac{(T_1 s + 1)(T_2 s + 1)}{(T_3 s + 1)(T_4 s + 1)}$$ (K1-26)		$-\frac{1}{T_1} \quad -\frac{1}{T_3} \quad -\frac{1}{T_2} \quad -\frac{1}{T_4}$	$T_1 = R_2 C_2$ $T_2 = (R_1 + R_3) C_1$ $T_3 T_4 = [(R_1 + R_3) R_2 + R_1 R_3] C_1 C_2$ $T_3 + T_4 = (R_1 + R_3) C_1 + (R_2 + R_3) C_2$

TABLE K-1 (continued)

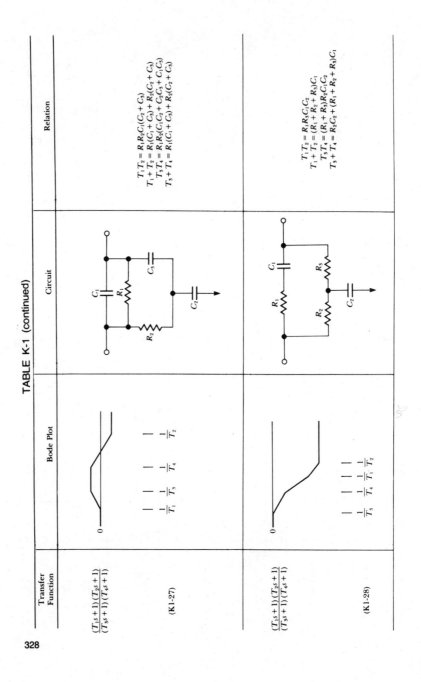

Transfer Function	Bode Plot	Circuit	Relation
$\dfrac{(T_1 s + 1)(T_2 s + 1)}{(T_3 s + 1)(T_4 s + 1)}$ (K1-27)			$T_1 T_2 = R_1 R_2 C_1 (C_2 + C_3)$ $T_1 + T_2 = R_1(C_1 + C_3) + R_2(C_2 + C_3)$ $T_3 T_4 = R_1 R_2 (C_1 C_2 + C_2 C_3 + C_1 C_3)$ $T_3 + T_4 = R_1(C_1 + C_3) + R_2(C_2 + C_3)$
$\dfrac{(T_1 s + 1)(T_2 s + 1)}{(T_3 s + 1)(T_4 s + 1)}$ (K1-28)			$T_1 T_2 = R_1 R_3 C_1 C_2$ $T_1 + T_2 = (R_1 + R_2 + R_3)C_1$ $T_3 T_4 = (R_1 + R_3)R_2 C_1 C_2$ $T_3 + T_4 = R_2 C_2 + (R_1 + R_2 + R_3)C_1$

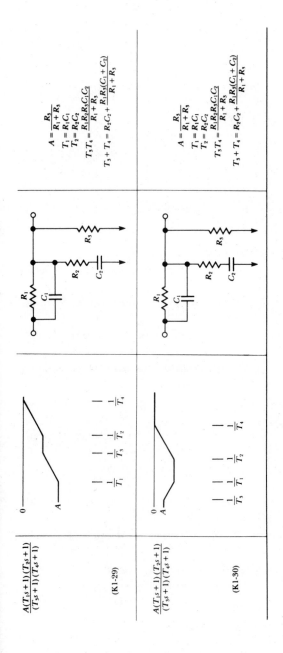

$$\frac{A(T_1s+1)(T_2s+1)}{(T_3s+1)(T_4s+1)}$$

(K1-29)

$$A = \frac{R_3}{R_1+R_3}$$
$$T_1 = R_1C_1$$
$$T_2 = R_2C_2$$
$$T_3T_4 = \frac{R_1R_2R_3C_1C_2}{R_1+R_3}$$
$$T_3+T_4 = R_2C_2 + \frac{R_1R_3(C_1+C_2)}{R_1+R_3}$$

$$\frac{A(T_1s+1)(T_2s+1)}{(T_3s+1)(T_4s+1)}$$

(K1-30)

$$A = \frac{R_3}{R_1+R_3}$$
$$T_1 = R_1C_1$$
$$T_2 = R_2C_2$$
$$T_3T_4 = \frac{R_1R_2R_3C_1C_2}{R_1+R_3}$$
$$T_3+T_4 = R_2C_2 + \frac{R_1R_3(C_1+C_2)}{R_1+R_3}$$

BIBLIOGRAPHY

Carlson, A. B., *Communications Systems*, 3d ed., McGraw-Hill, New York, 1985.

Close, C. M., *The Analysis of Linear Circuits*, Harcourt, Brace & World, New York, 1966.

Daryanani, Gobind, *Principles of Active Network Synthesis and Design*, Wiley, New York, 1976.

DiStefano, Stubberud, and Williams, *Feedback and Control Systems*, Schaum Outline Series, McGraw-Hill, New York, 1967.

Dorf, Richard C., *Modern Control Systems*, Addison-Wesley, Reading, Mass., 1967.

Edminister, J. A., *Electric Circuits*, 2d ed., Schaum Outline Series, McGraw-Hill, New York, 1983.

Gray, P. E., and Searle, C. L., *Electronic Principles*, Wiley, New York, 1969.

Hayt, W. H., Jr., *Engineering Electromagnetics*, 4th ed., McGraw-Hill, New York, 1981.

Hill, F. J., and Peterson, G. R., *Digital Systems: Hardware, Organization, and Design*, Wiley, New York, 1973.

——, and ——, *Switching Theory and Design Logic*, Wiley, New York, 1974.

Horowitz, P., and Hill, W., *The Art of Electronics*, Cambridge University Press, 1980.

330

Hostetter, Gene H., Savant, Clement J., Jr., and Stephani, Raymond T., *Design of Feedback Control Systems*, Holt, Rinehart, and Winston, New York, 1982.

ITT Staff, *Reference Data for Radio Engineers*, Howard W. Sams, Indianapolis, 1977.

Lowenberg, E. C., *Electronics Circuits*, Schaum Outline Series, McGraw-Hill, New York, 1967.

Malvino, A. P., *Electronic Principles*, 3d ed., McGraw-Hill, New York, 1984.

Melsa, J. L., and Schwartz, D. G., *Linear Control Systems*, McGraw-Hill, New York, 1969.

Millman, J., and Halkais, C. C., *Integrated Electronics*, McGraw-Hill, New York, 1972.

Roth, Charles H., Jr., *Fundamentals of Logic Design*, 3d ed., West Publishing Company, St. Paul, Minn., 1985.

Circuit Design, The Benjamin/Cummings Publishing Company, Inc., Menlo Park, Calif., 1987.

Schilling, D. L., and Belove, C., *Electronic Circuits: Discrete and Integrated*, 2d ed., McGraw-Hill, New York, 1979.

Spiegel, M. R., *Advanced Mathematics for Engineers and Scientists*, Schaum Outline Series, McGraw-Hill, New York, 1971.

———, *Fourier Analysis*, Schaum Outline Series, McGraw-Hill, New York, 1974.

Stremler, F. G., *Introduction to Communication Systems*, Addison-Wesley, Reading, Mass., 1977.

Tobey, G., Graeme, J. G., and Huelsman, L., *Operational Amplifiers: Design and Application*, McGraw-Hill, New York, 1971.

INDEX

Abbreviations and symbols, x
Ac (alternating-current), analysis,
    JFET, 277–279
Ac equivalent circuit, 4, 6, 9, 28,
    29, 45, 47, 54, 55, 69, 241,
    243, 245, 247, 248, 250, 252
Active filter, 11, 113–115
A/D (analog-to-digital) converter,
    219–222
AGC (automatic gain control)
    circuit, 10–13, 22, 41–43
Alpha (α), 31, 50, 148, 153, 156,
    229–230
Ammeter, dc, 177–180, 183–185
Amplifier:
    buffer, JFET, 20–21
    CB, 2–6, 8–11, 43–45, 243–244
    CBF, 61–63, 251–253
    CBR, 248–249
    CC, 2–5, 40, 246–247
    CD, 20–21, 259–261
    CE, 5–7, 19, 22–23, 52, 57–61,
        244–246
    CER, 22, 46–49, 249–251
    CG, 26–28, 258–259
    CS, 26–28, 32–33, 49–51, 53–
        57, 256–258
    CS/CE, 68–71
    differential pair, 2, 66
    feedback, 57–61, 61–64, 69–70
    gain control, 10–13, 22–23, 41–
        43
    h-parameter interrelationships,
        253–254

Amplifier (*Cont.*):
    hybrid-π, 51–53, 255
    IF, 235–238
    maximum power CC amplifier,
        39–41
    midband gain, 53–57
    Miller-effect analysis, 51–53
    multistage, 1–5, 26–28, 53–56,
        57–61, 65–71
    operational (*see* Operational
        amplifier problems)
    (*See also* Transistor)
Analog multiplier, 10–13
Analog simulation, 133–136
Analog-to-digital (A/D) converter,
    219–222
Analog-to-digital format, 220–222,
    233–234
AND gates, 201, 218
Angular position servo, 132–133
Antenna, transmitter, 230–233
Astable multivibrator, 74–76
Atmospheric attenuation, 231–232
Average value of a waveform,
    183–185

Background noise power, 233
Bandpass filter, 53, 98–103, 235
Bandwidth:
    ECAP 3-dB, 29–31
    filter, 98, 105, 118–122
    multipole, equations, 273–274
    multistage JFET, 53–57

Bandwidth (*Cont.*):
  operational amplifier circuit, 113
  receiver, 230, 238
BCD (binary-coded decimal) code,
  195
Bessel function, 238
Beta (β), 29–31, 50–51, 280–286
Biasing circuits, JFET, 277–286
Binary-coded decimal (BCD) code,
  195
Binary format (*see* Digital format)
Binary transmission, 233–234
Bipolar transistor equations, 241–
  255
Bode analysis:
  in CS three-stage amplifier, 56
  in determining K, ω_1, ω_2, and ϕ_m,
    136–138
  in finding $C(s)/R(s)$, 130
  in lag compensation, 155–160
  in lead compensation, 150–155
  in loop-delay analysis, 144–146
  in loop-gain function, 126–127
  in network synthesis, 98–103,
    113–115
  RC transfer function equations
    and, 319–329
  Routh check with, 130
  synthesizing $G(s)$, 138
  typical plots of transfer functions
    and, 289–297
  usage of, 125
  worksheet, 172
Boolean algebra, 201, 204, 222
Break frequency, 30, 53–57, 130,
  136, 138–140, 145, 151
Buffer, JFET, 20–23

Cancellation compensation, 147–
  150
Canonical form, 132, 135
Capacitor current waveform, 85
Capacitor energy, 79, 80, 105
Capacitor fall time, 96–98

Capacitor initial conditions, 88
Capacitor reactance, 2, 6, 52, 107,
  114–115
Capacitor rise time, 74–76, 83–84,
  103–105, 270–272
Capacitors:
  in RC circuits, 25, 29, 30, 78,
    83–84, 88–89, 96–97, 270–
    272
  in RC network synthesis, 302–
    329
  standard values for, 299
  (*See also RC* entries)
Carbon composition resistor
  standard values, 297
CB (common-base amplifier)
  equations, 2–6, 8–11, 43–45,
    243–244
CBF (collector-base feedback
  amplifier) equations, 61–63,
    251–253
CBR (common-base with base
  resistor amplifier) equations,
    248–249
CC (common-collector amplifier)
  equations, 2–5, 40, 246–247
CD (common-drain amplifier)
  equations, 20–21, 259–261
CE (common-emitter amplifier)
  equations, 5–7, 19, 22–23, 52,
    59–61, 244–246
CER (common-emitter with emitter
  resistor amplifier) equations,
    22, 46–49, 249–251
CG (common-gate amplifier)
  equations, 26–28, 256–259
Channel capacity, 233–234
Characteristic impedance, 229
Circuit:
  AGC, 10–13, 22, 41–43
  miscellaneous equations, 269–
    274
  RC, 25, 29, 30, 78, 83–84, 88–
    89, 96–97, 270–272
  RL, 17–20, 82, 270–271
  RLC, 88, 90, 95
  critical damping of, 95–97

Circuit analysis problems, 77–118

Clamping diode circuit, 23–25

Closed-loop transfer function, 130–132, 135, 142, 147, 149

Common-base amplifier (CB) equations, 2–6, 8–11, 43–45, 243–244

Common-base feedback amplifier (CBF) equations, 61–63, 251–253

Common-base with base resistor amplifier (CBR) equations, 248–249

Common-collector amplifier (CC) equations, 2–5, 40, 246–247

Common-drain amplifier (CD) equations, 20–21, 259–261

Common-emitter amplifier (CE) equations, 5–7, 19, 22–23, 52, 57–61, 244–246

Common-emitter with emitter resistor amplifier (CER) equations, 22, 46–49, 249–251

Common-gate amplifier (CG) equations, 26–28, 258–259

Common-source amplifier (CS) equations, 26–28, 32–33, 49–51, 53–57, 256–258

Communication link:
atmospheric, 230–234
hard-wire, 228–229
transmission line, 229–230

Compensation networks:
cancellation, 146–150
lag, 157–160
lead, 150–155

Component values, standard, 298–299

Control system (see Servo system)

Counting, synchronous, 190, 194, 213

Critical damping of RLC circuit, 95–97

CS (common-source amplifier) equations, 26–28, 32–33, 49–51, 53–57, 256–258

Current gain, 2–3, 5–7, 48–49, 62–64, 243–252

Current movement meter, 178, 179, 183

Current pulse applied to capacitor, 79

Current source, 66, 86–87

D flip-flop, 209–215

Damping, critical, of RLC circuit, 95–97

dB (decibel), 29–30, 53–57, 98–103, 113, 126–127, 131, 138–142, 144–146, 151–160, 229–232

dBm, 228–229

Dc (direct-current) ammeter, 178, 179, 183

Dc analysis:
CE, 3, 67–68
of diode, 37–39
of JFET, 21–23, 31–32, 41–43, 49–51, 277–286
of operational amplifier, 16–17, 33–37, 72–73, 178–179, 181–182, 223–224, 265–267
of transistor, 3, 21–23, 40–41, 67–68

Dc power supply, 185–186

Delay: loop, 144
loop-delay analysis, 144–145
phase, 144

Delta (Δ), 8, 10, 43–46, 47–48, 62–63, 245–253

Delta (δ), 238

Delta-Y transformation:
equations, 269–270
problems, 25–26, 150

Density, power, 230–232

Designation number, 195, 205–206

Deviation ratio, 238

Diagram, state, 211
Digital format:
    analog-to-, 220–222, 233–234
    AND gates, 192, 194, 201, 218
    BCD to Gray, 195–201
    Boolean algebra in, 185, 201,
        204
    D flip-flop in, 209–215
    designation number in, 195,
        205–206
    JK flip-flop, 189–191, 191–194
    Karnaugh map for, 196–198,
        203–204, 207–208, 211–212,
        216–217
    minterm representation in, 195
    NAND gate in, 199, 201–203,
        207–209, 213–222
    OR gates, 192, 194, 201–203
    pyramiding, 215
    standard basis in, 195
    state diagram for, 211
    synchronous counting in, 190,
        194, 213
    truth table with, 193, 195
Diode:
    clamping, 23–25
    dc analysis of, 37–39
    dynamic resistance, 22–23
    gain control with, 22–23
    rise time with, 83–84
    zener, 17–20, 24, 124, 179–180,
        267
Directional gain antenna, 230
Direct-current (DC) ammeter, 178,
    179, 183
Distortion:
    square-wave, 107–109
    triangular waveform, 227–228
Divider, analog, 10–13
Drain current equations, JFET,
    279–280
Dynamic resistance, 11–12, 21–22

ECAP 3-dB bandwidth, 29–30
Efficiency, telephone line, 228–229

Emitter follower (CC), 2–5, 40,
    246–247
Energy, capacitor, 78, 79, 105
Equations:
    bipolar transistor gain and
        impedance, 249–253
    diode, 37–39
    Fourier series waveform, 300–
        301
    JFET dc and ac parameters,
        277–287
    JFET small-signal, 256–261
    miscellaneous circuit, 269–274
    operational amplifier (see
        Operational amplifier
        equations)
    RC network synthesis for
        feedback amplifiers, 302–318
    RC transfer function, 319–329
    servo system, 126, 131, 132, 140
Eta (η), 229
Examination details, vi–ix

Feedback amplifiers, 57–64, 69–70
    RC network synthesis for, 302–
        318
Feedback problems, 126–175
Filter: bandpass, 53–57, 98–103,
        107–109, 118–122, 236–238
    high-pass, 105–107
    low-pass, 104–107, 108, 113–
        115
Flip-flop, D, 209–215
Flip-flop, JK, 191–194
Forcing functions, Laplace
    transform table of, 275–276
Fourier series:
    square-wave, 107–109
    triangular, 183–185
Fourier series waveform equations,
    300–301
Frequency:
    break, 30, 53–56, 130, 136, 144
    deviation ratio, 236
    geometric midpoint, 98

Frequency (*Cont.*):
  half-power, lower and upper,
    105–107
  oscillation, 74–76, 125
  transmitter, 230–233
Function:
  loop-gain, 127, 136, 144, 153,
    155–156, 160–170
  transfer (*see* Transfer functions)

g_{fs}, 20, 26–29, 279–280
g_{fso}, 20, 28, 279–280
Gain:
  current, 2–3, 5–7, 48–49, 62–
    64, 243–252
  voltage, 3–10, 22–23, 26–31,
    33–37, 46–48, 53–57, 62–64
Gain antenna, 230–233
Gain control, 10–13, 22, 41–43
Gamma (γ), 229–230
Geometric midpoint frequency, 98
Gray code, 195–201
Gunn oscillator, 235

h-parameter interrelationships,
  253–254
h_{ib} definition, 241
Half-power frequency, lower and
  upper, 105–107
Harmonics, square-wave, 107–109
Hartley-Shannon Law, 234
High-pass filter, 106
Hybrid-π amplifier, 51–53, 255

I_{DSS}, 12, 20–21, 26, 28, 31–32, 41–
  43, 49–50, 277–285
IF (intermediate-frequency)
  amplifier, 235, 237
Impedance:
  characteristic, 229
  input, 3, 4, 8, 26, 44, 48, 62, 70
    of current source, 86–88
    of voltage source, 86–88
  output, 8, 29, 43, 45

Impedance (*Cont.*):
  transfer, 57–61
  transfer impedance functions,
    302–318
Inductor:
  equations for, 270–271
  initial conditions for, 88
  protection circuit, 17–20
  RLC circuit with, 90, 95
  synthesized, operational
    amplifier, 13–15
Initial conditions, capacitor and
  inductor, 88
Input impedance (*see* Impedance,
  input)
Intermediate-frequency (IF)
  amplifier, 235, 237
Inverting operational amplifier
  equations, 264
Isotropic radiation, 231

JFET (junction field-effect
  transistor):
  AGC, 10–13, 22, 41–43
  biasing circuits, 280–286
  buffer, 20–23
  dc analysis, 20–21, 31–32, 49–
    51, 279–286
  divider, 10–13
  drain current equations, 279–
    280
  g_{fs}, 20, 26–29, 279–280
  g_{fso}, 20, 28, 279–280
  I_{DSS}, 12, 20–21, 26, 28, 31–32,
    41–43, 49–50, 277–285
  multiplier, 10–13
  pinch-off region analysis, 279–
    286
  pinch-off voltage, 21, 26, 28, 31,
    32, 41–43, 279–286
  r_{DS}, 10–13, 41–42, 279–287
  r_{DSO}, 42, 279, 287
  small-signal equations, 256–261
  terms defined, 278–279
  triode region analysis, 287
  V_G dependence on temperature,
    247

JFET (junction field effect
transistor) (*Cont.*):
voltage gain, 10–11, 26–28, 33,
69–70, 256–261
JK flip-flop, 191–194

Karnaugh map, 196–198, 203–
204, 207–208, 211–212, 216–
217

Lag, phase, 155–160
Lambda (λ), 229–230
Laplace transform table of forcing
functions, 275–278
Laplace transforms used in
problem solutions, 80–81, 83–
84, 88–89, 95–96, 103–105
Lead phase, 150–155
Logic (*see* Digital format)
Loop delay, 144
Loop-delay analysis, 144–146
Loop-gain function, 127, 136, 144,
153, 155–156, 160–170
Low-pass filter, 104, 105–108,
113–115

Margin, phase, 136, 142, 146, 151,
153, 155–157, 160–163, 170–
172, 175
Maximum power CC amplifier, 40
Maximum power transfer, 91–92,
110–112
Measuring system, stylus, 180
Metal-film resistor standard values,
296
Meter:
ammeter, dc, 177–180, 183–
185
current movement, 179–180
temperature, 177–183
Microwave transmitter, 234–238
Miller effect, 51–53
Minterm representation in digital
format, 195

Mixer, 234–238
Modulation sensitivity, 234–
238
Multiplier, analog, 10–13
Multipole bandwidth equations,
273–274
Multistage amplifier, 1–5, 26–29,
57–61, 65–71
Multivibrator, astable, 17

NAND gate, 199, 201–203, 207–
209, 213–222
Nepers, 230
Network equations, miscellaneous,
269–274
Network synthesis, 98–103, 113–
115, 118–122
Nichols diagram, 162, 175, 292–
297
Noise, background, 233–234
Noise effective temperature, 230
Noninverting operational amplifier
equations, 262–263
Nonlinear voltage regulator, 64–
65
Nyquist rate, 234

ω_π, 171, 173
ω_n, 143–144, 146, 153
Operational amplifier equations:
dc analysis of, 265–267
equivalent ac circuit of, 29–31
inverting, 264
noninverting, 262–263
offset analysis, 67–68, 267–268
temperature dependent, 71–74
Operational amplifier problems,
10–17, 25–26, 33–37, 39–41,
71–74, 113–115, 118–124,
134–136, 149–155, 159, 177–
183, 187, 222–225
OR gates, 201–203
Oscillation frequency, 74–76
Oscillator:
Gunn, 235
sinusoidal, 225–228

Output impedance, 8, 29, 43, 45
Overshoot, percent, 142–144

Partial fraction expansion, 80–81
Percent overshoot, 142–144
Phase delay, 144
Phase lag, 155–160
Phase lead, 150–155
Phase lock loop, 218–220
Phase margin, 136, 138, 142, 150–156, 160–163, 170–173
Phase plot, 137, 152, 157
Phi (ϕ), 136, 151, 155
Pinch-off region analysis, JFET, 279–286
Pinch-off voltage, 21, 26, 28, 31, 32, 41–43, 279–286
Polar plots of typical transfer functions, 292–297
Potentiometer gain constant, 132
Power density, 230–233
Power supply, dc, 185–186
Power transmitter, 230–233
Problems:
    circuit analysis, 77–124
    discrete semiconductor and integrated circuit, 1–76
    instrumentation, computer and systems, 177–238
    servo control and feedback, 125–175
Pyramiding, 215–218

Quantizing error, 221
Quantizing levels, 233–234

r_{DS}, 10–13, 41–42, 279, 287
r_{DSO}, 42, 279, 287
Radiation, isotropic, 230
Ratio, deviation, 238
RC circuit, 25, 29, 30, 78, 83–84, 88–89, 96–97, 271–272
    switch applied to, 103–105

RC network synthesis for feedback amplifiers, 302–318
RC transfer function equations, 319–329
Reactance, capacitor, 2, 6, 107, 114–115
Read-only memory (ROM), 227
Receiver bandwidth, 230, 238
Recorder, measuring system, 180–183
Regulator, nonlinear, 64–65
Resistor amplifiers:
    CBR equations, 248–249
    CER equations, 22, 46–49, 249–251
Resistor standard values:
    carbon composition, 299
    metal-film, 298
Rise time, capacitor, 75, 84–86, 103–105, 271
RL circuit equations, 17–20, 82–83, 270–271
RLC circuit, 88, 90, 95
    critical damping of, 95
    switch applied to, 88, 90
rms (root mean square) value of triangular waveform, 183–185
ROM (read-only memory), 227
Root-locus diagrams of typical transfer functions, 292–297
Routh criterion, 126, 129, 133, 142, 167–168

Semiconductors, discrete, problems dealing with, 1–76
Sensitivity, modulation, 236
Servo control and feedback problems, 125–175
Servo system:
    analog simulation of, 134
    angular position, 132–136
    block diagram of, 129, 130, 133, 135, 136, 144, 146, 151, 155, 160, 164, 168
    Bode (see Bode analysis)
    Bode worksheet, 172

Servo system (*Cont.*):
  cancellation compensation, 146–150
  canonical, 132, 135
  characteristic equation for, 126, 129, 133, 164
  compensation network (*see* Compensation networks)
  GH (j$\omega\pi$), 171, 172
  lag compensation, 155–160
  lead compensation, 150–155
  loop delay, 144–146
  loop gain, 57, 126–127, 136, 144, 156, 159, 160
  M_p, 160–163
  Nichols, 160–163
  Nichols worksheet, 174
  Nyquist, 168–173
  overshoot, 142–144
  phase margin, 136, 138, 142, 150–156, 160–163, 170–173
  polar plot worksheet, 173
  root-locus, 163–167
  Routh, 126, 129, 133, 142, 167–168
  settling time, 142–144
  stability analysis, 125–133, 136–143, 144–150, 160–172
  straight-line approximation, 128, 130, 137, 139, 141, 145, 152, 154, 157, 158
  velocity error constant, 155
Settling time, 142–144
Signal-to-noise ratio, 230, 233
Sinusoidal waveforms, 23–25
Square-wave input to circuit, 81–83, 107–109, 123
Stability analyses, 125–133, 136–150, 160–172
State diagram, 211
Statistical analyses (RSS), 222–225
Step voltage input, 75, 84, 88, 90, 95, 96
Stylus measuring system, 180–183
Superposition theorem, 93–95, 110–112
Switch:
  applied to *RC* circuit, 103

Switch (*Cont.*):
  applied to *RLC* circuit, 88–90
Symbols and abbreviations, x
Synchronous counting, 190, 194, 213
Synthesizing:
  active network, 122–124
  circuit to meet data, 115–118
  desired response, 113–115, 118–122, 122–124
  ECAP operational amplifier circuit, 29–31
  transfer functions, 81–83, 98–103, 105–107, 130

Tables:
  Bessel, 238
  capacitor standard values, 299
  carbon composition resistor standard values, 299
  Gray code, 195–201
  Laplace transform table of forcing functions, 275–276
  metal-film resistor standard values, 298
  RC transfer function equations, 319–329
  standard basis, 195
  transfer function plots for types of transfer functions, 289–297
  transfer impedance functions, 302–318
Tau (τ), 131–132, 140
Telephone line efficiency, 228–229
Temperature, noise effective, 230
Temperature-sensitive resistor, 177–183
Theta (θ), 148
Thevenin's theorem, 93–94, 109–113
Transducer, temperature, 177–179, 180–183
Transfer functions:
  closed-loop, 130–132, 135, 142, 147, 149
  filter, 81–83, 98–103, 105–107
  root-locus diagrams of typical, 289–297

Transfer functions (*Cont.*):
  synthesizing circuits, 81–83, 98–
    103, 105–107, 130
  typical plots of, 289–297
Transfer impedance functions,
    302–318
Transformer, 91–92, 228–229
Transformer equations, 273
Transistor:
  astable multivibrator, 17
  dc analysis, 3, 21–23, 31–32,
    41–43, 49–51, 67–68
  differential pair, 1–5, 66
  h-parameter interrelationships,
    253–254
  h_{ib} definition, 241
  hybrid-π, 51–53, 255
  input impedance (*see* Impedance,
    input)
  maximum power of, 39–41
  Miller effect, 51–53
  output impedance, 5–8, 29–31,
    46
  Thevenin's analysis, 93–94,
    109–113
  transfer function (*see* Transfer
    functions)
  transfer impedance, 57–61
  transfer impedance functions,
    268–284
  (*See also* Amplifier)
Transmission line, 229–230
Transmitter antenna, 230–233
Transmitter deviation, 238
Transmitter frequency, 230–233
Transmitters:
  microwave, 234–238
  power, 230–233
Trapezoidal voltage waveform, 85
Traveling-wave tube, 234–238
Triangular waveform, rms value,
    183–185

Triode region analysis, JFET, 287
Truth table, 193, 195

Up/down counter, 227

V_G dependence on temperature,
    280
Velocity error constant, 155
Voltage controlled oscillator, 220
Voltage gain, 10–11, 26–28, 33,
    256–261
Voltage regulator, nonlinear, 64–
    65
Voltage source, 86–87

Waveforms:
  average value, 183–185
  rms value, 183–185
  sinusoidal, 23–25
  square-wave, 80–81, 107–109
  step, 75, 84, 88, 90, 95, 96
  trapezoidal, 85
Worksheets, servo:
  Bode plot, 172
  Nichols chart, 174
  polar plot, 173

Xi (ξ), 142–144

Zener diode, 17–20, 24, 124, 179–
    180, 267